How Dangerous is Lightning?

Christian Bouquegneau
Vladimir Rakov

Edited by
William Beasley
School of Meteorology
University of Oklahoma

Dover Publications, Inc.
Mineola, New York

Copyright

Bibliographical Note

This Dover edition, first published in 2010, is the English translation of *Doit-on craindre la foudre?*, originally published by EDP Sciences, France, in 2006.

International Standard Book Number

ISBN-13: 978-0-486-47704-6
ISBN-10: 0-486-47704-5

Manufactured in the United States by Courier Corporation
47704501
www.doverpublications.com

Contents

Introduction

ENCOUNTERS of early humans with lightning undoubtedly were frightening and fascinating. All ancient civilizations incorporated lightning and thunder in their mythology (*see chapter 1*). Mythology and observations of lightning damage constituted lightning history till the Enlightenment period when Benjamin Franklin and Thomas-François Dalibard and his coworkers finally applied a scientific approach to this natural phenomenon (*see chapter 2*). However, scientists had to wait until the late nineteenth to early twentieth century, when modern instrumentation became available, to begin detailed studies of lightning.

What does science tell us about lightning today? The answer to this question can be found in chapters 3, 4 and 5, containing current views of thunderstorm clouds and their evolution, lightning phenomenology and parameters, spatial distribution of lightning activity, and the global electric circuit. Some other planets of the solar system also experience lightning, although quite different from its terrestrial counterpart. As of today, about 1,000 lightning discharges have been artificially initiated (triggered) from natural thunderclouds using small rockets extending thin wires between the thundercloud and Earth. This variety of lightning facilitated a substantial improvement in our knowledge of atmospheric electrical discharges.

In chapter 6, physical effects of lightning are reviewed. These are not mysterious any more. We will consider electrical, electrodynamic, thermal, electromagnetic, electrochemical, acoustic (thunder) effects, and, above all, lightning effects on humans and animals. In chapter 7, we introduce secondary effects of lightning and basic safety rules. We will also review types of damage due to lightning.

Lightning protection is a topic of great practical importance. The lightning attachment process is critical in this regard, but it is still one of the poorest understood lightning processes (*see chapter 8*). Presently, a simple electrogeometric model is used in designing lightning protection. Does one need to protect his/her house from lightning? This question is often asked and will be addressed in chapters 8 and 9. Lightning protection of power lines will be also discussed. The concept of risk and its practical evaluation will be introduced, based on the new international standard (IEC 62305) on lightning protection, which has been published in 2006.

Finally, in chapter 10, we look at the new frontiers in our understanding of lightning. We will discuss a variety of transient luminous phenomena occurring between cloud tops and the ionosphere, most of them in association with ordinary lightning. Recent observations of energetic radiation (X-rays and gamma-rays) from lightning will be reviewed. If the Earth's climate is indeed warming up, as predicted by many scientists, one should expect a significant increase in lightning activity in the years to come.

In order to make reading easier for the technical non-expert, a number of insets are included throughout the text introducing the basics of electrical phenomena. An appendix containing a more advanced background on electrical discharges in the air is intended for the reader who would be interested in strengthening his/her knowledge of the physics involved.

How Dangerous Is Lightning? is written for the general audience and, therefore, is accessible to everyone. It may be a good starting point for engineers, meteorologists, foresters, ecologists, architects, and physicians, who have to account for lightning and its hazards in their work. It is ideal for students working on their science projects. The original version of this book was published in French by EDP Sciences and contained 184 pages, including 47 illustrations. This version (in English) can be viewed as an updated version of the original one.

The authors are grateful to Pierre Lecomte and Frederic Coquelet for drawing most of the figures of this book and Marcus Moore and Sergei Rakov who helped with the typing.

Part I
Is Lightning Still Mysterious?

It is reasonable to infer from prehistoric artifacts that lightning has always fascinated human beings, so much so that it became an element of mythology and religion. Though lightning can be fascinating, it can be frightening as well, even to modern man, as anyone who has experienced a close cloud-to-ground strike will attest. Though many scientific mysteries about lightning still remain to be uncovered, we can certainly dispel the myths and explain much of the physics and phenomenology of lightning today on a scientific basis. For example, we know that lightning is of importance to the ozone balance in the atmosphere and that it serves to maintain the global electric circuit. Lightning resupplies the negative charge to Earth that otherwise would be lost within about ten minutes because of the conductivity of the atmosphere. Moreover, lightning may have played a crucial role in the origination of life on our planet. In 1924, Russian biochemist Alexander Oparin published his work on the origin of life. According to his hypothesis, lightning flashes provided the energy to synthesize prebiotic gases, which were present in the atmosphere a few billion years ago. In 1953, the young American chemist, Stanley Miller, who worked in Harold Urey's laboratory at the University of Chicago, generated a large number of high-voltage electrical discharges in a gas mixture composed of marsh gas, ammonia, hydrogen, and water, creating a number of amino acids and organic compounds typically found in biological systems. Among current hypotheses for the origin of life on Earth, including hydrothermalism (from the ocean floor) and extraterrestrial sources of interstellar dust produced in stars rich in carbon, Oparin's hypothesis and Miller/Urey's experiment continue to appeal to many scientists concerned with the origins of living things.

> Lightning may have played a crucial role in the origination of life on Earth.
>
> Lightning drives the global electric circuit, and also produces nitrogen oxide that supports plant growth.

Unfortunately, lightning also kills and destroys. It is very important that we know when and why lightning is a danger to avoid.

Sometimes it may be difficult to avoid the danger of lightning completely, but there are numerous precautions that can be taken to reduce the risk of injury and death and mitigate damage. There is, however, no method that would allow one to inhibit occurrence of lightning discharges from thunderclouds. Timely warning and adequate protection remain the only options to avoid deleterious lightning effects.

1
Mythology of Lightning

Beneficial or Harmful?

It is reasonable to speculate that in prehistoric times, lightning ignited dry tree branches, providing light and heat to the first humans, probably well before they learned to light a fire by themselves. According to numerous myths, only gods possessed fire. It is probable that the first tales and myths created by humans were inspired by natural phenomena, the meaning of which they could not grasp but wanted to interpret in order to allay their fears. If myths are similar regardless of the civilization that produced them, it is not likely because of communication among ancient civilizations, but because ancient man's thoughts originated from images of the world that are similar. People believed that lightning was a supernatural force. Everywhere in the world mythological stories appeared, taking into account the power of gods: punitive lightning, rumbling and terrifying thunder, lightning related to life-giving rain, or tamed lightning as a source of energy.

Ancient Mythology

In Asia Minor, Anatolian pantheons were led by a god of thunderstorms coming from the mountain, symbolized by a bull, and a goddess of fertility. The oldest representations of lightning gods are found on a cylindrical Akkadian seal in the Louvre from the first Babylonian period (2200 B.C.): a god who governs meteors, holding a whip, his cart is pulled by a mythical animal, and a female divinity holds the sky's fire.

In Hittite Anatolia, the thunder god is the great warrior Tarhunda-of-the-Sky, standing on a bull, armed with a mace. This commemorates the victory of the god of thunderstorms against the blind powers of the irrational. He corresponds to Adad in Semitic Mesopotamia, Ishkur in Sumer, Martu of the nomads of western Sumerians, Amurru of the Semites, the Hadad and Reshef of the western Semites, and Baal of the Canaanites (Syria) where he occupies the first rank in Ougarit Mythology. Baal is often represented as a young archer, armed with a mace and holding a young bull in leading strings. The bull is his fetish animal. All of these gods have the same origin and the same attributes as the Hurrit god Teshub: bull, lightning, mace, or mallet. As king of gods, Teshub has nothing left that is cosmic. He is the symbol of the human kingdom, the first paternal and patrician god, surrounded by a court and servants.

In Pharaonic Egypt, the Universe works by the contradictory actions of Osiris, who maintains the force of renewal in nature (vegetation, the Nile, the moon, and the sun) and Seth, his brother, malevolent power, violent and murderous, who destroys and manifests himself as god of rain and of thunderstorms, but also of desert and sterility. Each god is doubled with a feminine figure who, according to her companion's character, symbolizes maternity (Isis, sister and wife of Osiris) or sterility (Nephtys, sister and wife of Seth). Seth was considered to be identical to the Greek god Typhon, while Zeus' role was taken by Amon. Later, in Ptolemaic Egypt, Serapis combined the characteristics from both Osiris and Zeus returning to the supreme power of the latter.

In the Wonders and Fontanalbe Valleys (Mercantour Park) in Southern France, close to Nice and the Italian border, stands Mount Bego, the sacred mountain in the part of France, which is most often struck by lightning. Among numerous rupestral engravings with representations of lightning flashes, dating around two millennia B.C., there is a particular one called "the sorcerer": an anthropomorphic body holding a triangular dagger in each hand *(see color insert)*. This sorcerer is a god of lightning brandishing his flashes, as Enlil-Bel in Mesopotamia at the same period (circa 2150 B.C.).

Classical Mythology

In ancient Greece, lightning was Zeus' weapon. The locations struck by lightning were dedicated to him. In Rome, like other celestial gods, Jupiter *(see figure 1)* punished people with lightning.

In one of the myths of the sovereignty, Prometheus, the Titan, holds Zeus' power in check. Since the first men lived in nocturnal darkness and in the cold, Prometheus took pity on them. He stole the celestial fire and offered it to our ancestors who learned how to master nature's forces. He made them stronger, more intelligent, and more skillful.

FIGURE 1. JUPITER, AFTER A DRAWING BY MARIE, AGE 10.

Zeus, Master of the Universe, didn't appreciate this and decided to punish Prometheus cruelly: he tied him up to a rock on top of a mountain. Each day a giant eagle lacerated his belly with its claws, and each night the wounds healed. And to punish men, Zeus sent them a dangerous trap, which nobody could escape—Pandora. She opened up the box in which all ills and diseases were locked. From then on, mankind has been condemned to aging and death. The lightning-fertility association, negative among the Greeks, is on the contrary positive in some other mythologies.

The lightning bolt, divine scepter, celestial super-weapon, manifestation of the divine wrath of an anthropomorphic deity with a powerful hand, bears various names: thunderbolt in ancient Greece, forged for Zeus by Hephaistos, the sky-dwelling blacksmith; bundle of lightning flashes; vajra (lightning flash or thunderbolt), belonging to Indra in the Indian subcontinent, or its analog dordje in Tibet, with a diamond's purity, symbol of stability in Buddhism; mjöllnir, the famous hammer of Thor in Scandinavia or of Donar among the Germans; an axe or even a double axe of Shango in Central Africa, among the Yorubas; thunderstone, mace (Tarhunda-of-the-Sky), bludgeon, mallet; sling; Categuil (or Illapa) among the Incas roamed the heavens with his sling and his mace shining with lightning flashes, creating so much damage on Earth that children were sacrificed in order to calm his wrath.

Vedic Mythology

From the top of his white elephant, or of a tricephalic elephant, Indra, the lightning striker, the Hindu god of storms, king of the heaven, used to strike the inhabitants of the Indian subcontinent with his dangerous weapon, the vajra. He was a benevolent and rough deity, bellicose but compassionate. He dealt with human affairs and was the *deus ex machina* who recognized heroes of talent deserving rewards.

In Indra's prophecies, one finds cosmogony, recipes for happiness and the science of predicting. Indra penetrated the Hindu pantheon by killing Vritra, snake of dryness, which had drunk the cosmic waters and then rested wrapped around mountains.

Indra's vajra opened the snake's stomach and the waters that escaped gave birth to life and freed the day's dawn. Lightning at the origin of life, it is more than just a myth!

In the fifth century B.C., the role of Indra, offering life on Earth, was taken by Vishnu. Indra was put out of favor because the devotees of Shiva and Devi dominated in the cities destroyed by the adherents of Indra's worship.

From Vikings to Gallic Mythology

Among the Vikings, the super-weapon was called mjöllnir, the famous hammer of Thor (*see figure 2*) or of Donar, then of Wodan, among the Germans. In other civilizations, one also finds the mallet (of the good god Sucellus in Gaul), the mace and the bludgeon (the god Tarhunda-of-the-Sky of the Hittites and the druid-god Dagda in Ireland).

Thor (god of Thunder), holding his hammer mjöllnir, had supernatural strength. He was also a benevolent god, a favorite of Scandinavians, because he protected men against evil, dispensed the rain and ruled over fertility. A dangerous warrior, he exterminated giants without fear. His red beard is plaited, he had a frightening voice, and his eyes would throw lightning flashes. Like a boomerang, his hammer had the power of coming back into his hand after striking. When thunder roams, it is Thor's chariot, pulled by he-goats, that rolls on the vault of heaven. When lightning strikes the ground, it is Thor who threw his weapon. Thor's hammer is the most common symbol found engraved on stones bearing runic inscriptions, or melted in beautiful current Nordic jewels. Thor died heroically in Ragnarök (the paradise), after a great fight against his hereditary enemy, the cosmic snake Jormungand, who threatened Earth by embracing it. Thor broke its head with one fatal blow then was

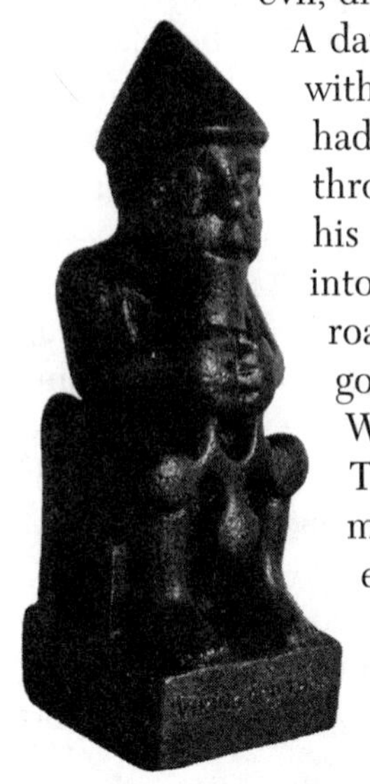

FIGURE 2. THOR (ICELANDER STATUETTE)

drowned by the torrent of venom, which came out of the beast's open jaws.

Is the legend of Thor and Jormungand much different from the one of Indra and Vritra? Why should it be? Indo-European mythologies, like Indo-European languages and cultures in general, follow the same schemes.

In Anglo-Scandinavian languages, Thor gave his name to Thursday. It is *torsdag* in Swedish, Danish and Norwegian, *torsdai* in Finnish, like Donar (his German analog coming from Donner) gave *donnerstag* in German and *donderdag* in Dutch, like Jupiter (*Jovis* in Latin) gave *jeudi* in French, *jueves* in Spanish, *giovedi* in Italian and *joï* in Romanian, the same Indo-European origin. Thursday is dedicated to the god of lightning.

In Scandinavia and among the Slavic people, Perun is worshipped and earns the same attributes as Thor. The Russian word for lightning, *molniya*, is apparently derived from the Vikings' mjöllnir. In Baltic countries, Perkunas (in Lithuania) or Perkons (in Latvia) would be the anthropomorphization of a tree, which embodies the fertile life, and thus the benefits of a particular worship.

Zeus in Greece, Jupiter in Rome, Indra in India, Thor and Perun in northern, central and eastern Europe have similar characteristics and attributes. They are gods who create and organize. Their resemblances are striking: they all brandish a throwable weapon, they possess a violent and mysterious strength which leads them to heroic acts, they fulminate, punish, reward and hold the secrets of wisdom. They also share the same weaknesses: they cheat and procreate everywhere. Both Zeus and Jupiter had numerous mistresses, both goddesses and mortals.

But the hierarchical idea remains preponderant: the world's reign belongs by right to the gods of Heaven, never to Earth or sea deities, even in the maritime nations like the Greeks or the Vikings.

Lightning is a weapon of the gods! In Judeo-Christian mythology, isn't it with lightning's complicity that Yahve came down to Mount Sinai to dictate the Ten Commandments to

Moses, another Prometheus, herald, mediator, fire bearer? In the Apocalypse accounts, at the great Last Judgement, lightning frequently appears as symbol of God's punishment.

Gallic ancestors worshipped Taranis (thunder in Gallic), personification of the luminous sky and thunderstorms, assimilated with the Roman Jupiter. His emblem is the wheel symbolizing the rolling of thunder, similar to the sound of the wheel on Roman roads. He is holding S-shaped hooks, symbols of the winding trail of thunderbolts, and is often displayed on top of or vanquishing a monster, victory of the sky over the Earth, of light over dark, of good over evil, and, why not, of civilization over barbarism. Traces of his cult can be found not only in Gaul but also in Great Britain, Germany, Hungary, and Croatia. Statues of Taranis are lodged at the top of pillars or columns that can be found in Romanized Gaul.

Protecting Saints

In western Europe, during a thunderstorm, an old custom encouraged the peasants to have in their pocket a thunderstone while reciting: "Peter, Peter, protect me from thunder." But many saints other than Peter were worshipped. There are more than twenty of them. Prayers were seen as actions that can appease a god's wrath.

There are seventy-six different Saint Donats, among whom Saint Donat, the bishop of Numidia in the fourth century, said to be from Münstereifel (Germany), is portrayed holding a bundle of fire in one hand and in the other a few ears of grain, which remind of his power in protecting the harvest against hail.

In 1652, the relics of Saint Donat taken from Saint Agnes's catacombs in Rome arrived in Euskirchen, three leagues away from Bad Münstereifel, their final destination, under unusually intense rain. The day after, on the day of Transfiguration, a miracle happened. During the mass celebrated by the Jesuit Father Herde, sent from Münstereifel to Euskirchen to prepare the pilgrimage, a terrible thunderstorm occurred. Thunder rumbled during the elevation and the communion and, before the last gospel, a thunderbolt penetrated into the church and struck the Jesuit father who was invoking Saint Donat (*see figure 3*). Against all odds (but this is a scientifically explainable fact), he survived and the same day managed to lead the triumphal entrance of the relics into Münstereifel. The invocation to Saint Donat was consecrated and he has been prayed to as saint protector against lightning ever since.

Examples of priests and faithful people struck by lightning are common. This fact suggested to Camille Flammarion to treat lightning, if not unbeliever or anti-religious, at least not respecting sacred places. So many superstitions!

Around the world, the most invoked saint in the Catholic Church is Saint Barbara, virgin and martyr. Daughter of the Syrian satrap Dioscore, Barbara was born approximately in the year 306. She was converted to Christianity against her father's will and was thrown in jail in a tower and submitted to the most atrocious tortures. Since she refused to renounce her faith, Dioscore ran out of patience and brought her in front of Judge Marcien who condemned her to death. Her father beheaded her himself. While coming down the mountain where he killed her, the father died of a violent lightning strike.

This is why Saint Barbara is associated with noise and fire. Removed from the Catholic calendar since 1969, she is still celebrated on December 4 as patroness of miners, artillerymen, artisans, and technicians!

She is represented holding a tower and is often accompanied by Saint Claire, both associated with the same prayer.

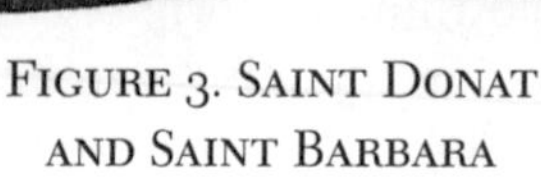

FIGURE 3. SAINT DONAT AND SAINT BARBARA

American Mythology

In the southwestern deserts of North America, the great show of a sky sparkling with repeated lightning strikes seems extraterrestrial. Under such a sky, the Native Americans changed their fear into a myth, the one of a spirit such as Ahayuta, god of lightning and god of war, among the Zunis in New Mexico, or the myth of a winged god, a thunderbird, such as Amoncas among the Kwakiutls on the Canadian Pacific Coast. This thunderbird (*see color insert*) is an eagle carved on totem poles. The flutter of wings produces thunder, which rumbles and rolls in the sky. The thunderbolts spring up out of sparkling eyes, as they sprang out of the only sparkling eye of each of the three Cyclops. The eagle is so strong that it can lift a whale in its terrifying claws. Nevertheless the thunderbird has no maleficent spirit, because, though its influence covers lightning and storms, it brings water, which gives birth to forests and plains.

The thunderbird is also found in ancient Mesopotamia (Zu), Siberia, Peru, and Mexico, as well as among the contemporary Zulu and Baziza people in South Africa where "to be struck by lightning" means "to be lacerated by the thunderbird's claws." The fierce eagle eats fish greedily and dies immediately, and its corpse is used to obtain ingredients for local medicine.

According to the mythology of the Dakota Sioux of North America, four thunderbirds fought the Earth god for the control of water. The big bird from the west, gifted with a holy supernatural power, Wakan Tanka, called its brothers. The sky had to be their realm. They won over the Earth god and reigned over water and fire, over thunder and wind, over life and death.

In the Amerindians, in pre-Columbian times, lightning was personified by a mythical being dressed with splendor and wearing a multicolored feather headdress. Tlaloc (Aztecs), Cocijo (Zapotecs), Aktsin (Totonacs), Tzahui (Mixtecs), Illapa (Incas), and Chac (Mayas) are all gods of lightning, rain, and fertility.

Peruvian farmers were aware of Illapa's activities; they beseeched him to provide enough water and they offered him big human immolations in case of long-lasting drought. There, lightning is also bound to divination, since Inca diviners held their gift because they had been struck by lightning.

Chac raises up lightning with stone axes and produces rain by throwing down calabashes full of water. After a long-lasting drought, Mayas set themselves far from their camp and prayed, fasted, and practiced sexual abstinence.

Thunder and flash of light are associated with the same lightning phenomenon. It is thus not surprising to find sparkling eyes, screams and wing fluttering of the thunderbird. Thunder, acoustic signature of lightning, is the ensemble of acoustical refractions, from the nearby blast wave to the muffled hum coming from a faraway strike. Sometimes, thunder resembles the terrifying roar of the jaguar or the celestial puma.

Among the Desanas (or wind's sons) from the deep and humid forests in Colombia (Amazonia), the thunder crash is attributed to the hoot of the owl, a nocturnal predator with silent wing flutters, messenger of death. Here, lightning is not a weapon held by gods, but an essential part of a big energy cycle, the raw material of life and magic, fertility materialized in palpable light, a seed coming from the sun, the semen that brings life to Earth!

African Mythology

Among the Yorubas, in Nigeria and Benin (sub-Saharan Africa), the great warrior Shango governs thanks to thunderbolts. His priests hold Shango thunder axe and believers bring stone axes on thunderbolt zigzag sticks as offerings to this god.

There, lightning animist priestesses brandish an *oshe*, Shango emblem, a symbol of dignity of his priestesses (*see figure 4*). It is a double-headed axe, stylized in wood, whose handle is simple or most generally carved in the shape of a female body, naked, symbol of ritual purity, and kneeling down, respecting both gods and kings.

Shango sanctuaries are decorated with women statues, statues of mother and child. Shango presides at births. His attributes are lightning, winds and rain, but also fecundity, the germinating power of plants, and the struggle with contagious diseases. Shango is virile and vigorous, violent, and a lover of justice.

He punishes liars, thieves, and wrongdoers. Being struck by lightning is a particularly defamatory way of dying. A house

FIGURE 4. OSHE SHANGO AND LIGHTNING PRIESTESS IN BENIN

struck by lightning is a house marked with Shango wrath. The owner will have to pay huge fines to this god's priests and will have to make offerings to appease him. The blood of animals which are sacrificed to him (often rams whose head blows have the suddenness of thunderbolt) is poured into the *erun ara*, the thunderstones thrown out by Shango like meteorites to maintain their strength and vitality.

More than other gods, Shango has marked immigrants with his print, since he is found in Brazil (Shango is the fourth god acting in the famous Candomblé in Bahia) and in the Caribbean Sea, namely in Cuba and in Haiti. In southern Haiti, Shango controls the stormy heavens; in northern Haiti he is syncretized with Saint John the Baptist, a man so violent that in order to reduce his power, God makes him drunk on his celebration day, the 24th of June, in the Caribbean, during the thundery summer. In Cuba, in spite of his virility, Shango is syncretized with

Saint Barbara (Santa Barbara), who reminds of loud things. In Brazil, Shango is now syncretized with Saint Jerome.

Ouidah is a nice sandy place in the Benin Gulf. There the Door of No Return (*see figure* 5) stands close to the harbor where African slaves embarked on long and deadly voyages to Central and South America in the eighteenth century. On the foot of this huge monument you can find sculptures of Shango as well as sculptures of Xevioso, his counterpart among the Fons. Xevioso, deity of lightning, looks like a ram, runs his fury in the clouds, vomits his axe, and throws lightning strikes. Xevioso is also a war god with a dog head or human head, as in Zwenge, a small Benin village, where you can find him paired with another fertility god, named Goun, civilizer hero who provides life-giving rain. Shango's double-headed axe is both destroying and protecting, in a duality of opposite energies (death/life) that is found in Indra's vajra or in Thor's mjöllnir as well.

The axe is simple among the Fons. Among the Bambaras, the demiurge (creator deity) Faro uses a whip instead, but the simple axe is also found among the Dogons and the Bambaras in Mali. The axe is thrown out from the sky to Earth by the

FIGURE 5. DOOR OF NO RETURN AND XEVIOSO AT OUIDAH (BENIN)

god of water and fertility. Stone axes or thunderstones are kept in the sanctuaries of these gods. They are ritually used to fight drought or are planted with seeds to improve their germination. These thunderstones can attract or repel lightning. Hung up on a house roof, they repel lightning. Conversely, set up under a shelter in the middle of the bush, they attract lightning. This example resembles the behavior of Gallic warriors who erected long spears vertically along the rivers and laid down on the ground nearby, believing that they would be protected against lightning.

Among the Fangs in Gabon, a thunderstone is placed between the legs of pregnant women to facilitate the delivery. What a strange parallelism! In Yakutia, women giving childbirth drink small pieces of thunderstones with water in order, they think, to get rid of the placenta more easily. The centerpiece thunderstone is a vaginal symbol.

Some years ago, one of the authors had the privilege of dining in Harare with Dr. Sibanda, Secretary of Legal Affairs and Culture of Zimbabwe, an impressive personality who told him wonderful stories about African village shamans who have great powers with their calabashes. Even today people go to shamans to counter danger or to cast a spell on a rival or an enemy, or even to make him disappear. He also claimed he was always successful. In other African tribes, lightning is personified in a magic bird (Umpundulo among the Basutos) or a snake (Ambelema among Bakango pygmies), as in ancient Japan the snake was considered as the thunder god Susanoo (Izumo region), a god who is sometimes malicious, sometimes beneficent.

Asian and Oceanian Mythology

In Chinese Taoist mythology, lightning is actually tamed. It is the consequence of the conflict between Yang (masculine symbol, warm air) and Yin (feminine symbol, cold rain). The most important deity who presides over the mysteries of lightning is Lei-Tsu. He chairs the Ministry of Thunder and Thunderstorms, made of twenty-four dignitaries among whom most famous are, besides Lei-Tsu himself, his associate Lei-Kung, Prince of Thunder, a blue winged man, horrible, with fearsome claws, who punishes humans for their most secret crimes. But Lei-Kung can only create thunder, rain being the creation of Yu-Tzu, Master of the Rain. Clouds are created by Yun-T'ung, Little Boy of the Clouds, wind is created by Feng-Po, Count of Wind, thunderbolts are made by Mother Lightning Tien Mu, who reflects them to human beings, with either destructive or constructive effects, thanks to the mirrors which she holds. Yi-King associates thunder to fear and war. Thunder, as disturbance of world and nature, comes from the breaking of yang and yin.

In Japan, in Kyoto, the Temple of the 1001 Buddhas (Sanju-Sangendo) is surrounded by thirty representations of spirits, among which is Raijin (or Raiden), god of thunder, who shakes Heaven and Earth with his striking hits (*see color insert*). The purpose was to instigate terror by showing the tremendous power of lightning gods.

The Nepalo–Tibetan Buddhist tradition includes intense worship of bodhisattvas (Buddha's to come) who, powerful and compassionate, offer help and protection to humans. In one of his hands, Vajrapani holds a dordje (or vajra, active and masculine element), lightning striking virile symbol of active compassion (symbol of stability and method in Buddhism), and in the other a ritual drilbu (or ghanta), ritual handbell, feminine symbol of wisdom, reminder of impermanence. Harmony and enlightenment come from the union of impermanent wisdom and active compassion. Numerous bodhisattvas hold this lightning-striking sceptre, including Vajrasattva, Vajradhara, and Vajrapan. As a divine instrument, the Nepalo-Tibetan dordje represents method, as opposed to wisdom and knowledge, symbolized by the drilbu (small bell, feminine and passive element) with sound resembling thunder (*see color insert*)

In March 1994, the sixty-five-mile-high central tower of the Angkor Vat temple in Cambodia was struck by lightning. One of the two co-prime ministers of the time took off right away to the site in order to preside at a religious ceremony. There was very little damage to the structure of the temple, which is eight centuries old, but this natural phenomenon was interpreted as a very bad foreboding that had to be countered with a series of propitiatory rites. In Cambodia, Angkor is a symbol of power, of greatness, of immortality. If you visit this enchanting site, you will also discover Ta Keo temple, with five towers on a pyramidal basis. This temple was never finished and its walls are deprived of any ornaments, because it was struck by lightning during its construction.

Oceania is just as rich in lightning mythology. In Arnhem Land on the north-central Australian Coast, are some of the oldest rupestral paintings with lightning motives (*see color insert*). The Aborigines of this region tell the story of the giant Djambuwal who guarded the beaches against newcomers. He wielded a large spear named Larrapan, but invaders from Borneo defeated Djambuwal and, as he was dying, he swore that they would hear his voice during each thunderstorm. The Larrapan sword was sometimes seen crossing the Australian sky like a shooting star. From time to time it bounced back on the rocks, generating sparks. In Arnhem Land, lightning is less a weapon than a spurting of thunderbolts, an energy source.

Farther west, the Gunwinggu people have a more sinister legend. Namarragon, the Lightning Man, wandered the skies holding a spear in each hand. During the rainy season, he came out of the sea and lived in the clouds. But he was more than a weather phenomenon. With his thundery voice, he addressed crowds. If some people sinned, his voice whistled fiercely and his swords pierced the sky with sparkling arcs, provoking trees and the ground to burst out, killing sinners. Namarragon incinerates the unfaithful wife and her adulterous lover in the hell of the heavenly fire.

These two versions of the same myth correspond to the climatic differences between tropical lightning (raging thunderstorms of the Gunwinggu) and the more temperate, less "aggressive" lightning (Arnhem Land). It is not surprising that in warm and wet areas (equatorial and tropical) people consider lightning as a terrifying phenomenon. In higher latitudes, lightning raises interest and it is not feared that much. After all, for the northern people, Thor chose lightning as his weapon (his hammer mjöllnir), while he could as well have chosen any other attribute.

Among the Maoris of New-Zealand, Tane, an anthropomorphic god, separates his parents Rangi (the sky) and Papa (the Earth) to create the world, by using lightning power, the supernatural fire (ahi tipua). Tawhaki is the mythic god of thunder that you can find everywhere in Polynesia. He wanders in the sky and exhibits his power, throwing thunderbolts out of his armpits. Nevertheless there are several shapes of flashes: flat flashes belong to the Lightning Lady (Hine-te-uira) and forked flashes to the Lightning Lord (Tama-te-uira). Thunder (whaitiri) also splits, though the Thunder Lady's voice (Hine-whaitiri) is more often heard during the various ceremonies.

All Over the World

Lightning has given birth to many myths from civilization's earliest times on. Myths are an appeasing projection of human anxiety. Through these concepts, profound demands, aspirations and men's dreams are organized. Mythological stories describe behaviors that persist through time. Identical patterns in thought, carved in the collective unconscious, build, by immanence, the sacred to which human beings constantly refer.

2
Historical Overview

Ancient Civilizations

As early as 1500 B.C., Etruscans observed nature and perhaps were the first to suggest that sharp points appeared to attract lightning. Around their temples, Egyptian priests built tall masts with copper lamellas, which can be considered as the first lightning rods. Under thunderstorms, mysterious "aigrettes" burst up from the tips, terrorizing the faithful who believed in the gods. These emanations, partial discharges in the air, now known as *corona* (*see appendix*) puzzled people because they were unable to explain them. Fifteenth-century navigators called them *Saint Elmo's fire*. They observed them on top of boat masts and considered them as a protection sign by Saint Elmo.

In 60 B.C., the Latin poet and philosopher Lucretius tried to explain this phenomenon without invoking the power of the gods. Though he was not a physicist, he talked about *original atoms of lightning* which entered matter in liquid fires. He was then very close to modern scientific concepts. He translated Aristotle's statement according to which a lightning flash runs much faster than the sound of thunder.

From ancient times on, fearless spirits tried to react in their own manner against lightning. Julius Caesar wore a laurel crown on his head to protect himself against it, because Pliny the Elder said that lightning never strikes laurels. As for Roman emperors Augustus and Severus Alexander, they used to crawl under a dead cow during thunderstorms.

In his *History of Rome*, Titus Livius quotes seventy-three examples of lightning damage on the highest buildings in the city and deduces that Jupiter is angry because of human arrogance. The Roman Senate decided that each lightning strike is to be declared. They prescribed repentance, prayers, and sacrifices for each damaged building.

Ancient people involved lightning in many aspects of private and public life. On Julius Caesar's death lightning struck close to the Capitol. The sky showed its wrath and announced Brutus' defeat. More recently, the Freedom genius on top of Bastille column in Paris was struck by lightning on April 8, 1866. This event was considered a sign of the high trials France was going to experience.

From the Middle Ages to the Enlightenment

The concept of natural calamity did not exist in the Middle Ages, at least in Western Europe. The Catholic Church dominated society with its ideology, imposing beliefs and behavior. It considered lightning as a warning or even a punishment for an individual or collective fault. According to the church, lightning only struck usurpers, dancers, fornicators, and blasphemers. In the famous *exempla*, gloomy stories of that time, a man having fornicated on Easter Sunday who received the sacrament the next day without any confession in the meantime was struck by lightning three days later. When a thunderstorm was approaching, church bells were rung to ward off lightning. Saint Thomas Aquinas declared "the tones of the consecrated metal repel the demon and avert storm and lightning." This was surely not effective. In Germany, for example, during one period of just thirty-five years, 386 churches were struck by lightning and more than one hundred bell ringers were killed.

Till the end of the seventeenth century, namely the time of French philosopher René Descartes, famous for his Method's Treatise, most explanations of lightning were quite incorrect. Indeed, the science of electricity is relatively young, having been largely ignored before the Enlightenment!

Circa 600 B.C., Thales of Millet noted that yellow amber attracts bits of straw, when amber is rubbed with fur. Then not until 1600 A.D. did William Gilbert (1540–1603) explain the attractive electric power of amber (*elektron* in Greek) and name this property of a material "electricity."

The quantity of electricity stored by a rubbed material is called *electric charge*. The electric charge generates an *electric force*, which, though well described quantitatively, is still not completely understood at a fundamental level. If we do not know yet what an electric charge is intrinsically, we pretty well know its effects.

ELECTRIC CHARGE

The electric charge is defined from the electric current.

The electric charge unit in the SI system (International System) is *the coulomb* (C); one coulomb is the charge transferred by an electric current of one *ampere* (A) during one *second* (s).

Experiments show that the total electric charge in an isolated medium is always conserved and remains constant.

On a spherical metallic conductor, electric charges distribute themselves uniformly on the exterior surface.

If the conductor has a non-spherical shape, electric charges concentrate on the exterior surface in regions where the radius of curvature is the smallest (the wonderful effect of points).

This effect is well known to physicists working on electrical discharges in gases (*see appendix*).

Indeed, it was not until 1660 A.D. that the mayor of Magdeburg (Germany), Otto von Guericke (1602–1686), who was famous thanks to his experiment of joined hemispheres pulled by horses to show effects of the atmospheric pressure, built the first machine to generate electric charges.

In 1729, Stephen Gray analyzed the differences between electric conductors and insulating materials (dielectrics). Studying the interactions between electrically charged materials, the French chemist Charles-François de Cisternay du Fay (1698–1739) noted in 1734 that some charges attract one another, while others repel each other. Thus he discovered existence of two types of electricity. He also designed an electrometer with elder spheres and golden sheets.

Curiosity rooms (*cabinets de curiosité* in French) were scientific lounges famous in France during the eighteenth century.

One of them was the *cabinet* of Abbé Jean-Antoine Nollet (1700–1770) who performed numerous electrostatic experiments and disseminated the information. The American, Benjamin Franklin (1706–1790) discovered atmospheric electricity, designed the first lightning rod (in 1752) and arbitrarily named the two types of electricity: *positive* or vitreous electricity (rubbing of glass on silk) and *negative* or resinous electricity (rubbing of wax on fur). The law deduced from du Fay's observations is: *like charges repel one another, while unlike charges attract one another.*

ELECTRIC FORCE

As for gravitational forces, an electric force F between two point charges q_1 and q_2 varies inversely proportionally to the square of the separation distance r.

Coulomb experimentally found that the electric force is proportional to the product of the charges q_1 and q_2.

The electric force acts along the straight line linking both centers of charges.

Like charges repel one another, unlike charges attract one another (*see below*).

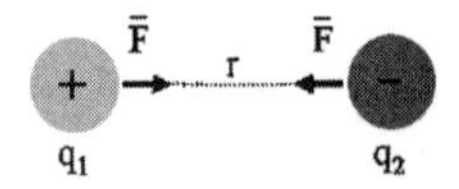

Coulomb's Law states that

$$F = k\frac{q_1 q_2}{r^2}.$$

where k is proportionality constant to be determined.

The proportionality constant k is defined taking into account the units chosen and characteristics of the medium surrounding the charges.

In free space or in the air (where k is smaller only by 0.06 %)

$$k = 9 \times 10^9 \text{ Nm}^2/\text{C}^2.$$

Charles de Coulomb (1736–1806) quantized the electric charge and the attractive or repulsive electric force, making electricity a new quantitative science, at the end of the eighteenth century. In 1745, at Leyden (the Netherlands), Petrus van Musschenbroek (1692–1761) designed the first capacitor to store large amounts of electric charge: the famous *Leyden jar*. Franklin later accomplished the same goal with a parallel-plate capacitor.

During these times, electric sparks were generated artificially in the laboratory (in 1746, by Johann Heinrich Winkler, professor of classical philology, at the University of Leipzig, Germany) and in nature (in 1747, by Marquess Scipione de Maffei in Verone, who had the intuition of upward discharges).

ELECTRIC INDUCTION AND GROUNDING

If a neutral conductor A is grounded by means of a wire and we bring close to it negatively charged material B, by electric induction, the electrons from conductor A (grounded) move as far as possible from the negatively charged material B. The electrons spread out into the ground, leaving the conductor A with a positive charge.

If the material B is moved far from conductor A, electrons come back from the ground and neutralize the positive charge on conductor A.

The Earth is a huge reservoir of negative electric charges and has the capacity to store a large amount of charge.

Every conductor in contact with the ground is said to be earthed or grounded, which serves to free it from any excess free electric charge, as described above.

Franklin or Dalibard?

Thanks to his prodigious intuitive genius, Benjamin Franklin discovered atmospheric electricity. He believed that there was only one kind of electricity and reasoned that electric "fluid" was redistributed when it moved from one material to another. The material with a deficit of electric charges was electrified negatively and the material with an excess of electric charges was electrified positively. This implied the new name of "positive" for vitreous electricity and "negative" for resinous electricity. Franklin also designed the first lightning rod, after his famous kite experiment, probably carried out in June 1752. He succeeded to charge a Leyden jar and proved that electric charges were present in a thundercloud. Let us stress that, according to Franklin and the physicists of that time, the lightning rod was supposed to protect against lightning by progressively discharging the thundercloud and thus avoiding any further electrical discharge. We presently know that this is not how the lightning rod works.

In his kite experiment, Franklin had risked his life, but we realized this only later on. Making sparks flowing between his hand and a metallic key linked to a wet hemp string, he was very lucky to survive (*see figure* 6). Indeed, he would have died in the case of a direct lightning strike to the kite. Another fearless scientist, Professor Georg Wilhem Richmann, working in Russia, installed an ungrounded iron rod having connection with his experimentation room and was killed by a direct lightning strike to the rod.

FIGURE 6. FRANKLIN'S KITE EXPERIMENT
(AMERICAN PHILOSOPHICAL SOCIETY LIBRARY, PHILADELPHIA)

On the roof of a Philadelphian house, Franklin installed a metallic rod, which could "neutralize" a lightning discharge, by means of its bottom end being driven into the ground. Franklin's lightning rod was born. It spread throughout America. More than ten thousand rods were installed in less than ten years. Its efficiency was immediately confirmed, particularly on church steeples—preferential targets for lightning. For example, the Campanile of Saint Marco Basilica in Venice was several times damaged or completely destroyed by lightning. After installing a Franklin lightning rod in 1766, it never again experienced any lightning damage.

But church people went on fearing that their faith, not being afraid any more by the sky fire, will loose respect for God's creation. Just the opposite occurred: the more scientists enriched their knowledge on lightning, the more they admired nature's power. Fear was warded off and mystery remained!

In France, as early as May 1752, Thomas-François Dalibard was asked by the Comte de Buffon to translate Franklin's book entitled *Experiments and Observations on Electricity*. Dalibard prepared the famous experiment of Marly-la-Ville on the northern outskirts of Paris (*see figure* 7). During a thunderstorm, Coiffier, his assistant, drew four-centimeter-long sparks. In contrast with the grounded lightning rod, the Marly vertical rod was isolated from ground. As with most other experiments of that time, the Marly experiment only served to prove the electrical nature of lightning, not to protect any structure against it.

Lightning became the newest fashion. Atmospheric electricity was experienced in physics "cabinets." These were attended not only by scientists, but also by clergymen and noblemen, as well as elegant ladies. In his cabinet, Abbé Nollet installed an isolated lightning rod. He gave private lectures on experimental physics in front of King Louis XV who enjoyed seeing a chain of people shocked when Nollet discharged a Leyden jar through their hands. He also experienced the "electric kiss," generating a spark between the lips of a couple of young people with a replica of the first Otto von Guericke electrostatic generator.

In spite of his divine inspiration in his study of the relation between lightning discharges and laboratory electric sparks, Abbé Nollet was completely mistaken in his "fumy" theory

FIGURE 7. MARLY-LA-VILLE EXPERIMENT, ACCORDING TO DALIBARD

of "effluence and affluence." He was very jealous of Franklin, a brilliant self-taught genius, who overshadowed the French school.

A year after Franklin's discoveries, a magistrate of Nérac (Gascony, France) named Jacques de Romas, seemed to ignore American discoveries. He repeated Franklin's experiments using a 240-meter-long kite string, surrounded with a conducting material all along its length to improve electric conduction. He succeeded in generating twenty-centimeter-long sparks from this string and then, with longer and longer strings, electric arcs, up to three meters long in the open air.

In 1753, Gianni Battista Beccaria, professor of experimental physics at Torino University (Italy), supported Franklin's theory and denounced manipulation and interpretation errors committed by Nollet who had haughtily attacked Franklin. Moreover, he was the first to imagine sending a small rocket to the thundercloud, with a metallic wire connected to the ground, so as to withdraw electricity from the cloud. This nice idea became extremely important in the second half of the twentieth century, when small rockets were launched to trigger lightning discharges. These experiments were first conducted over water using ships (namely the Thunderbolt!) off the Florida Coast.

In his essay on nature's interpretation (1753), French philosopher Denis Diderot (1713–1784) noted the great revolution in science, characterized by the new experimental philosophy where those who think and those who act meet together in front of the authority of the facts. By the end of the eighteenth century, all educated people who could afford it hurried to erect a lightning rod on their buildings.

In France, people went further. They used *lightning umbrellas* with a metallic wire touching the ground as well as *lightning hats* for the elegant Parisian ladies—another ephemeral fashion.

The Saint-Omer Trial

Outside Paris, people did not follow this fashion. The Saint-Omer trial in the northern part of France is proof. In 1780, an old lawyer who respected physics, named Charles Dominique de Vissery de Bois-Valé erected a lightning rod on his roof. His neighbors worried, thinking that God would punish this blasphemy. They sued him and won. De Vissery appealed to the higher court of Artois in Arras. He was defended by a brilliant young lawyer named Maximilien de Robespierre and secured the right to install the lightning rod. Winning this trial in a masterly manner, Robespierre convincingly linked the law and science. He said that truths are founded on particular facts, independent of any theory. This exemplary trial with many repercussions (it lasted until May 1783) was the trial of enlightenment against superstition. De Vissery won and the young provincial lawyer began on the path to his eventual reputation. This is an abstract of Robespierre's speech for the defense:

Arts and sciences are the richest gift that the sky made to human beings. From which fatality did they find so many obstacles to get set up on earth? Why cannot we pay to famous men, who invented them or led them to perfection, the right tribute of gratefulness and admiration that the total humankind owe them, without being obliged to lament on these shameful persecutions? Let misfortune reach the one who enlightens his fellow-citizens! Ignorance, prejudices and passions have shaped a terrible league against ingenious men to punish the services they will offer to their fellows.

This fantastic episode, illustrating the victory against superstitions at the end of the eighteenth century (the lightning rod being considered as a scientific application, was the symbol against obscurantism and of the victory of human genius over natural forces), was unfortunately not followed by many significant discoveries regarding lightning during the nineteenth century, except for the first photographs of flashes and ingenious measurements of stroke duration, often less than one millisecond.

Modern Era

In the nineteenth century, a lot of statistical data was collected on the location and frequency of lightning discharges. In France, this data was collected by the Ministry of Justice. The famous French book *The Caprices of Lightning*, by Camille Flammarion, was published in 1905. This is a short abstract evoking the author's writing with some humor:

In 1791, a young farmer girl stood in a meadow during a thunderstorm, when all of a sudden a fireball appeared at her feet, caressed her naked feet, flew under her clothes, got out of her bodice and burst to open air noisily. At the time the fireball penetrated under the girl's petticoat, the latter broke open like an umbrella. The girl fell down backwards. Two witnesses ran to help her. But the young girl was not hurt. After a medical check up, physicians saw superficial erosion from her right knee to the middle part of her breast. Her shirt was destroyed at the corresponding trajectory and a small hole pierced from side to side.

These facts are real, not legends. Flammarion's book describes spectacular damage, including pulverized or fused objects, exploded trees, and burnt or petrified human beings.

Serious scientific investigations of lightning started primarily at the beginning of the twentieth century, after the invention of the streak camera by Vernon Boys in 1926. These cameras made it possible to record multiple-stroke flashes in South Africa and at the Empire State Building in New York.

At that time, power transmission lines in the USA, Germany, and Russia were equipped with magnetic links (pieces of ferromagnetic material), which have the property to get magnetized by intense electric currents flowing nearby. These magnetic links allow estimation of amplitudes of lightning currents striking power lines. Many towers and chimneys were also equipped with such links.

Thanks to cathode-ray oscillographs, scientists in Russia, England, and the USA could finally visualize the various shapes and characteristics of the lightning current components.

The famous Swiss scientist, Karl Berger worked during his whole professional life (1940s through 1970s) to observe and study the lightning electrical discharges at his experimental station on Mount San Salvatore, near Lugano, Switzerland. Berger's data on lightning parameters are still used as the primary reference source for both lightning research and lightning protection.

Now we reach the scientific part of our book!

How complete is our knowledge of lightning at the beginning of the twenty-first century? We simulate atmospheric discharges of up to ten million volts in high-voltage laboratories. Many researchers record and analyze flashes at experimental stations all over the world (USA, Brazil, Japan, China, and Europe). Lightning is presently triggered artificially from natural thunderclouds with small rockets trailing grounded wires at Camp Blanding in Florida and at Langmuir Laboratory in New Mexico, as well as at two sites in China. Global and regional observations of lightning are performed using satellites and various lightning locating systems.

Scientists recently discovered thunderstorm-related transient luminous phenomena above storms (sprites, elves, blue jets) and energetic radiation (X-rays, gamma-radiation) produced by lightning. Now we are in the heart of contemporary research.

Part II
What Does Science Tell Us?

As we have seen, research on lightning is as old as research on electricity. From the middle of the eighteenth century, this natural phenomenon preoccupied researchers from various fields of study. Two centuries later, the Scottish physicist Charles T. R. Wilson (1869–1959), a Nobel Prize winner who invented the cloud chamber, determined the amount of electrical charge inside a thundercloud by making measurements of the electric field (*see definition of the electric field in chapter* 3).

In this part, we explore the science of lightning in the following three chapters:

In chapter 3, we analyze the atmospheric phenomena, which lead to the generation of lightning flashes. The electrically charged cloud called cumulonimbus, king of the clouds, is the main source. This thundercloud seldom appears isolated. It can produce intracloud discharges, intercloud (cloud-to-cloud) discharges, cloud-to-air discharges, and cloud-to-ground discharges.

In chapter 4, we examine the cloud-to-ground discharge (lightning flash to ground) with emphasis on the various characteristics of the corresponding currents. We also introduce the concept of global electric circuit.

In chapter 5, we are concerned with lightning location and detection problems, as well as the triggering of lightning flashes and their presence on other planets of our solar system.

3

What Is the Source of Lightning?

Characteristics of Cumulonimbus

The thundercloud is the most vigorous among all types of clouds. It is known as the *cumulonimbus* or king of the clouds. Generally, *cumulonimbi* do not appear in isolation (*see figure* 8), but as cloud aggregates. They differ from other rain clouds by their larger dimensions, especially in the vertical direction, and by their ability to produce lightning.

An isolated cumulonimbus has the shape of an enormous vertical tower surmounted by an upper part of anvil shape, called the anvil, at an altitude between 6 and 17 km, sometimes even more. *Cumulonimbi* are observed in the troposphere where temperature decreases with altitude, up to the tropopause, the upper boundary of the troposphere.

Regions of Earth's Atmosphere

The tropopause altitude varies from about 10 km at the poles, to over 17 km in tropical regions. The stratosphere (*see figure* 8) extends above the tropopause. It is a region in which, after a slight decrease followed by a constant value (up to an altitude of 25 km), the temperature increases because of absorption of solar radiation by ozone, up to the stratopause, the upper limit of the stratosphere at about 45 km. In the mesosphere beyond the stratopause, the temperature decreases again up to a minimum at approximately 85 km, at the level of the mesopause. Together, the mesosphere and the stratosphere are often called the middle atmosphere. Beyond the mesopause, the temperature continuously increases (it reaches 1000 K at 750 km) in the thermosphere.

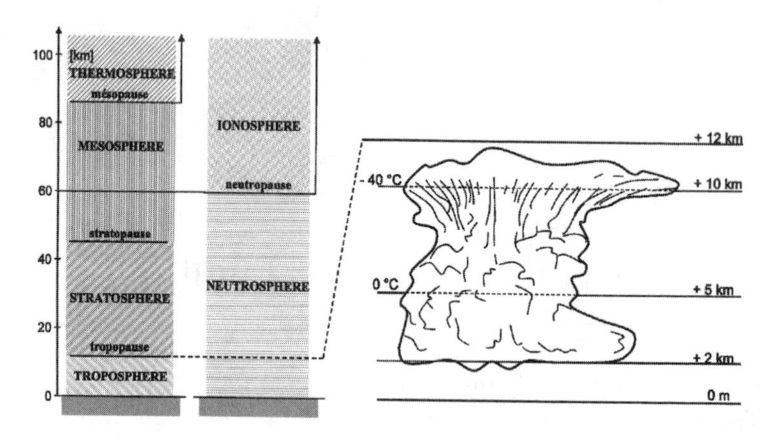

FIGURE 8. ATMOSPHERIC LAYERS, ELECTRIC REGIONS AND TYPICAL CUMULONIMBUS

From the point of view of global electric circuit, one needs to distinguish two regions of the earth's atmosphere: a lower region, the neutrosphere, where the electrical conductivity is low. It extends up to an altitude of 60 km or so, corresponding to the neutropause. Above the neutropause (from 60 to 500 km), the upper region is called the ionosphere or, more precisely, electrosphere, where the electrical conductivity is high.

The cumulonimbus height is related to the altitude of the tropopause. It varies with latitude and with the seasons. It can reach 17 km or more in tropical regions. In temperate regions (middle latitudes), it decreases from 12 km in summer to 6 km in winter. The cumulonimbus base is generally situated between 1 and 2 km above ground level and has a diameter of 10 km or so.

Evolution of a Thundercloud

The vigorous vertical growth of cumulonimbus clouds is a result of unstable masses of warm, humid air and large decreases in temperature with height over a large range of altitudes.

In the convective development phase, the isolated thundercloud, warmer than the ambient air, grows rapidly in the vertical direction. When water vapor condenses, the release of latent heat of condensation increases the instability of the original air mass increasing the rate of vertical growth. Updrafts force the cloud to reach higher regions where the temperature is well below freezing, the 0°C isotherm (line of constant temperature) being found typically at an altitude between 4 and 5 km. Inside the anvil, the updraft speed can reach thirty meters per second.

Radar echoes show liquid water drops and solid ice crystals that are lifted to high altitudes by the intense updrafts. When the liquid and solid water accumulations are such that the updraft cannot overcome their weight, the precipitation particles begin to fall. The mature phase of the cumulonimbus begins with the first precipitation. Rainfall increases beneath the cloud, carrying cold air with it to create a "gust" front. Such a thunderstorm cell usually contains hundreds of thousands of tons of water.

Updrafts and downdrafts coexist, but downdrafts finally win the competition, marking the beginning of the dissipation phase. The thundercloud pours out its last precipitation. The cloud breaks up by evaporation following the adiabatic (without any heat transfer) rewarming of the downdrafts caused by increasing pressure at lower altitudes. Sometimes, the cloud leaves fragments of cirrus clouds in place of the anvil. The complete life cycle of a *cumulonimbus* from first formation of small *cumulus* clouds to dissipation generally lasts about an hour.

Heat, or more precisely, a steep temperature decrease with height and humidity, even weak, has to be present in order for a *cumulonimbus* to form. In regions where one of these two factors is almost always absent, polar regions or in desert regions for example, lightning occurs very rarely.

Inside the thundercloud, positive and negative electric charges separate to form a gigantic electric dipole, or tripole. These electric charges are produced by collisions between the small ice crystals and the large (millimeter size) soft hail particles (graupel) in the presence of small water droplets. The sign of charge acquired by ice crystals and the graupel depends on the ambient temperature and the liquid water content. Small ice crystals are lifted by updrafts and large graupel particles are pulled down by gravity. A relatively strong electric field results.

When the electric field exceeds the dielectric strength of air (*see figure* 9), a lightning discharge is initiated inside the cloud.

ELECTRIC FIELD LINES

The concept of electric field or electric force field is defined in terms of its effects. A field of electric force exists in a spatial region when an electric charge, situated at any point of this region, experiences an electric force. Let a positive electric charge source +q be uniformly distributed over a small spherical surface. Let us study its influence on another positive test charge $+q_0$. The test charge is repelled by a radial force (*see chapter* 2).

Many different atmospheric conditions can lead to lightning. *Cumulonimbi* in summer are typical, but lightning can also

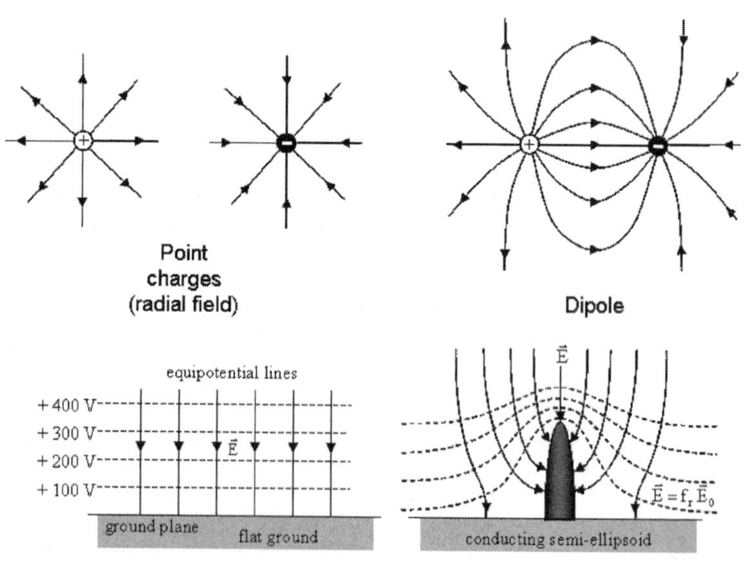

FIGURE 9.
ELECTRIC FIELD LINES GENERATED BY SOME TYPES OF SOURCES: POSITIVE POINT CHARGE, NEGATIVE POINT CHARGE, ELECTRIC DIPOLE, FAIR-WEATHER ELECTRIC FIELD OVER FLAT GROUND AND IN THE PRESENCE OF CONDUCTING SEMI-ELLIPSOID.

DEFINITION OF THE ELECTRIC FIELD

The concept of electric field lines of force was introduced by Michael Faraday (1791–1867) in 1812. The electric force experienced by a positive test charge at any point is oriented along the tangent to the line of force or electric field line at this point.

With a negative electric charge source, the radial lines of force are oriented in the direction of the charge source, attractive force (*see figure* 9).

The density of the lines of force decreases proportionally to the spherical surface ($S = 4 \pi r^2$); it is proportional to $1/r^2$. The electric field is proportional to the concentration of the lines of force. The more lines are concentrated, the stronger the electric field. The concept of lines of force guides our intuition about how charges behave in electric fields, but it does not say anything about the nature of the electric field.

The electric field E at any point is the electric force F, experienced at this point by a positive test charge $+q_0$, divided by the magnitude of the charge:

$$E = \frac{F}{q_0} = k\frac{q}{r^2} = 9\ 10^9\frac{q}{r^2}.$$

The unit of electric field is the volt per meter (V/m) in the SI. The electric lines of force are often called electric field lines. The electric field lines originate from a positive charge and terminate on a negative charge. Electric field lines never cross.

Inside a conductor, there is no electric field ($E = 0$). The electric field is perpendicular to the exterior surface of the conductor, either when the electric field is created by the charged conductor itself or by induced charges on the initially neutral conductor placed in an external electric field. This property allows one to isolate, electrically speaking, any object by surrounding it completely with a *Faraday cage*, i.e., a conductive enclosure, inside which no electric field exists.

occur during hurricanes and in snowstorms in winter. Lightning often occurs in mesoscale convective systems that spread out horizontally over hundreds of kilometers. Lightning usually does not occur when clouds do not contain ice or when updrafts are not sufficiently strong.

Electric Charge Distribution

As shown in figure 10, the charge distribution in a mature *cumulonimbus* is often approximated as an electric tripole, with a positive charge of 10 C to 50 C, sometimes up to 300 C in the upper part of the cloud, negative charge of similar magnitude in the middle, typically at the level of the −10°C isotherm, separated from the positive upper charge by a quasi neutral region, and a small pocket of positive charge typically between 1 and 5 C, (sometimes a little more) situated in the bottom part of the cloud, below the 0°C isotherm.

DIELECTRIC STRENGTH

The dielectric strength of a medium in a uniform electric field is the critical value of this field at which electrical breakdown occurs in this medium, making it conductive.

In air at sea level, the dielectric strength in a uniform field is 30 kV/cm (kilovolts per centimeter) or 3 MV/m (megavolts per meter or millions of volts per meter).

Any protrusion at ground level leads to an increase of the local electric field. On top of a sharp ellipsoid, where the ratio of the large axis to the small axis is 30, the electric field is strengthened by a factor of 300 (*see figure 9, bottom right*). Since the dielectric strength in a uniform field is 3 MV/m, an ambient electric field 10 kV/m (kilovolts per meter) is sufficient to initiate corona on its top and possibly an upward leader (*cf. appendix*).

Classification of Lightning Discharges

The most often encountered type of electrical discharge is the intracloud discharge, i.e., a discharge inside the *cumulonimbus*

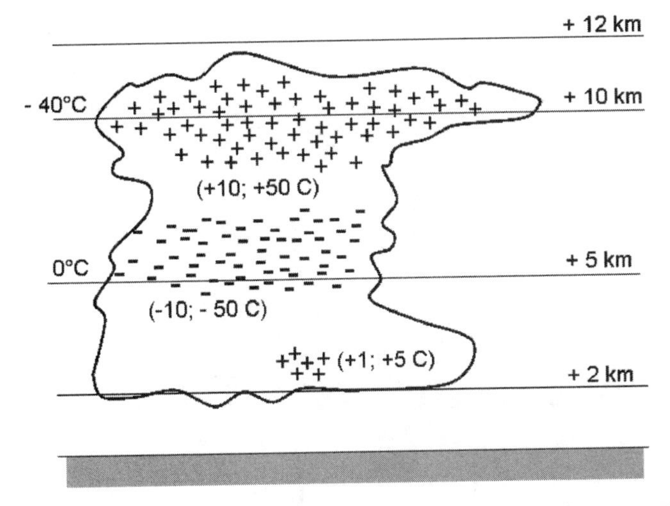

FIGURE 10. DISTRIBUTION OF THE ELECTRIC CHARGES IN AN ISOLATED CUMULONIMBUS (TRIPOLE)

itself (*see figure 11.1*). More rarely, a discharge may propagate from within the cloud into the ambient air outside the cloud (*see figure 11.2*). These are called air discharges. Sometimes, very rarely, the discharge starts from a region of one *cumulo-*

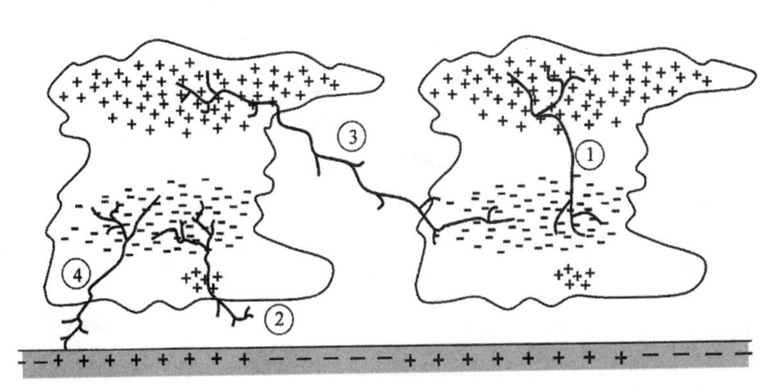

FIGURE 11. DIFFERENT TYPES OF LIGHTNING DISCHARGES:

1) INTRACLOUD DISCHARGE;

2) AIR DISCHARGE (CLOUD-TO-AIR);

3) INTERCLOUD OR CLOUD-TO-CLOUD DISCHARGE;

4) CLOUD-TO-GROUND DISCHARGE

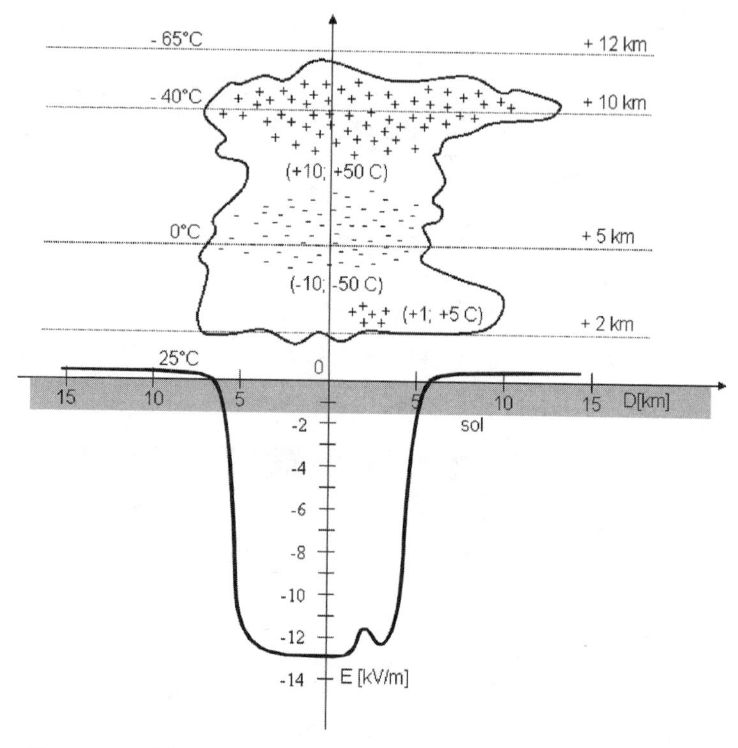

FIGURE 12. ELECTRIC FIELD E CREATED BY A CUMULONIMBUS AT GROUND LEVEL

nimbus and extends to a region of opposite charge in another *cumulonimbus* (*see figure 11.3*). Such discharges are called intercloud or cloud-to-cloud discharges. Intracloud, cloud-to-air, and intercloud discharges are referred to as in-cloud or just cloud discharges, often abbreviated as IC. Cloud-to-ground (CG) discharges, less frequent than many people think, are most often initiated from the lower negative charge distribution (*see figure 11.4*), but can also start from the upper or lower positive charge.

In summer, only about 10% of CG discharges are positive. Negative discharges are most common. In winter and under some meteorological conditions, positive discharges may occur more often than negative during part or all of the lifetime of a storm. The ratio of CG discharges to the total number of discharges (CG + IC) increases with increasing latitude. As an average on a worldwide scale, it appears that there is one CG discharge for every three to ten IC discharges.

Recently, different types of discharges in low-density air well above cloud tops have been discovered. We will describe these transient luminous events in a later chapter.

Electric Field Change at the Ground

In its mature phase, the potential difference between the ground and the negative charge region in a *cumulonimbus* can be as much as 100 MV (megavolts). The electric field at the surface beneath the cloud can become very large. Figure 12 shows the electric field beneath a typical *cumulonimbus* at

ELECTRIC POTENTIAL

Around a source charge q which creates an electric field in its vicinity, let us bring a test charge q_0. The electric potential energy, W (measured in joules), of the test charge is proportional to this test charge. When dividing the potential energy by q_0, you obtain the electric potential energy per unit charge or electric potential U or simply potential:

$$U = \frac{W}{q_0}.$$

The electric potential unit is the volt (V).

One volt is equal to one joule per coulomb (1 V = 1 J/C). The potential U at a distance r from the source charge q is

$$U = k\frac{q}{r} = 9\ 10^9 \frac{q}{r}.$$

As an example, let us consider an isolated spherical conductor of radius R = 1 m with +1 μC (microcoulomb) electric charge. One finds $U_R \cong 9$ kV and $E_R \cong 9$ kV/m.

If the conductor has no spherical symmetry, but contains some regions with smaller radii of curvature (sharp tips, for example), the electric field is strongly enhanced (*see figure 9, bottom right*).

ground level. In flat ground areas and fair-weather conditions (without thunderstorms), the vertical component of electric field is 100 V/m or so, because of the presence of negative charge on earth and an equal amount of positive charge distributed in the atmosphere (often assumed to reside in the electrosphere or ionosphere). When a *cumulonimbus* forms or appears overhead, the electric field at ground level reverses its polarity and increases typically by a factor of 10 or 100. Usually the field at the surface is limited by corona to about 10 kV/m (ten thousand volts per meter).

General Reference

V.A. Rakov and M.A. Uman, *Lightning – Physics and Effects*, Cambridge Univ. Press, 2003.

4
Cloud-to-Ground Flashes

General Information

The term "flash" when used to refer to lightning is susceptible to many different interpretations. In its simplest use, it refers to the optical manifestation of lightning, although "flash" and "lightning" are often used interchangeably. The principal discharge consists of plasma, i.e., an ionized medium in which individual atoms and molecules have net charge, but, macroscopically, the positive and negative charges are equal. This medium is considered a fourth state of matter, different from solid, liquid, and gas. Ordinary flashes are sometimes referred to as linear. Sometimes lightning channels appear to be beaded. Such flashes are called bead lightning.

Thunder can seldom be heard from flashes more than about 25 km distant. Flashes without audible thunder are sometimes referred to as "heat" lightning.

Though there are more than three thousand papers on ball lightning, there is no consensus yet on the mechanism(s) responsible for this phenomenon. Numerous ball lightning observations, in the shape of bright spheres associated with cloud-to-ground strikes, leave no doubt as to their existence, but these spheres have never been unambiguously photographed or reproduced in the laboratory.

Karl Berger, already cited in chapter 2, proved the random nature of lightning strikes in a given region at different times, but he never observed ball lightning. He considered ball lightning to be an optical illusion due to the fact that our eyes are dazzled or films are overexposed in case of an intense lightning flash. Clearly, more observations are needed to understand the phenomenon of ball lightning.

Karl Berger also showed that lightning does not necessarily always strike the highest points, but often much lower targets, namely hillsides. Since Berger's time these facts have been confirmed by numerous objective observers and researchers.

The tortuosity of lightning discharges remains puzzling. It probably results from random changes in direction of successive steps, but that is a description, not an explanation. Tortuosity of lightning discharges occurs on scales from more than 1 km to less than 1 m. Thanks to studies of triggered lightning (*see next chapter*), tortuosity on very short scales (less than 10 cm long) has been observed.

The first electrical discharges in a storm occur inside the thundercloud, sometimes a quarter of an hour before the first cloud-to-ground flash. For lightning warning (in the areas of aerial navigation, telecommunications, transport and distribution of electrical energy), this quarter of an hour can provide a safety margin.

Classification of Cloud-to-Ground Flashes

Let us now consider negative cloud-to-ground discharges or negative CG flashes.

A cloud-to-ground discharge is typically initiated by a preliminary discharge inside the thundercloud. According to Karl Berger, four types of cloud-to-ground discharges can occur (*see figure 13*), depending on the polarity and direction of propagation of the stepped leader. This stepped leader is not very luminous, and in any case it is overwhelmed by the bright flash that occurs after it forms, so it cannot be discerned by the naked eye. If we take into account a third condition of having a complete discharge or a discharge stopping in midair, there are a total of eight possible situations.

In flat ground areas, the most usual type is the *downward negative leader* (90% in summer), or, more rarely, the *downward positive leader* (especially in winter and in some types of severe storms). However, from taller structures, an *upward*

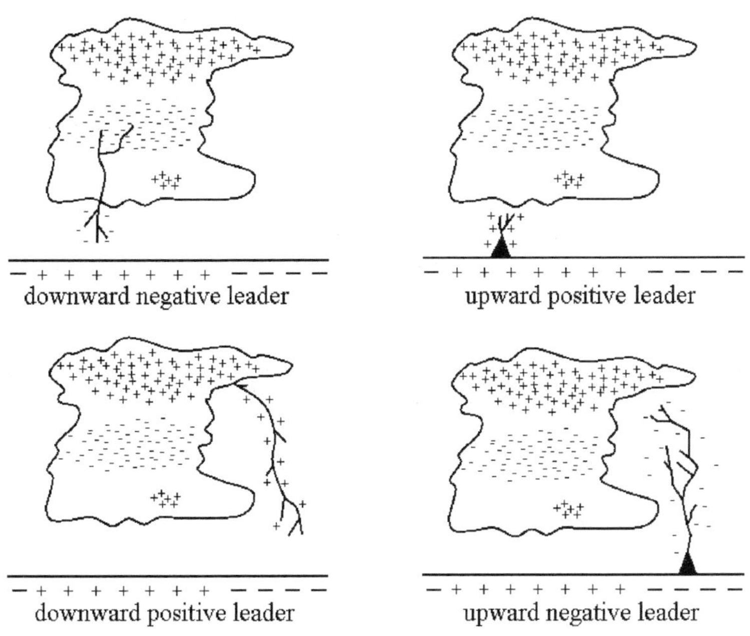

FIGURE 13. FOUR TYPES OF CLOUD-TO-GROUND DISCHARGES

positive leader or, less often, an *upward negative leader* can occur. There are other approaches to lightning classification.

Cloud-to-Ground Flash Phenomenology: Flat Ground

Let us describe the mechanism of a negative flash initiated by a *downward negative leader* in the case of flat ground. Figure 14 shows development (as would be seen using a streak camera) of a stepped leader initiating a three-stroke negative cloud-to-ground discharge or negative flash.

A downward leader propagates step by step (steps of 10 to 200 m) at a relatively low velocity (100 km/s or so) with 1 kA (one thousand amperes) peak current pulses, formed during 1 μs (one microsecond or one millionth of a second), the time intervals between steps being 20 to 50 μs. When the leader tip approaches ground, the electric field strongly increases and initiates positive upward connecting leaders from sharp objects or ground irregularities. This gives rise to the attachment process. The length of upward connecting leader is of the order of

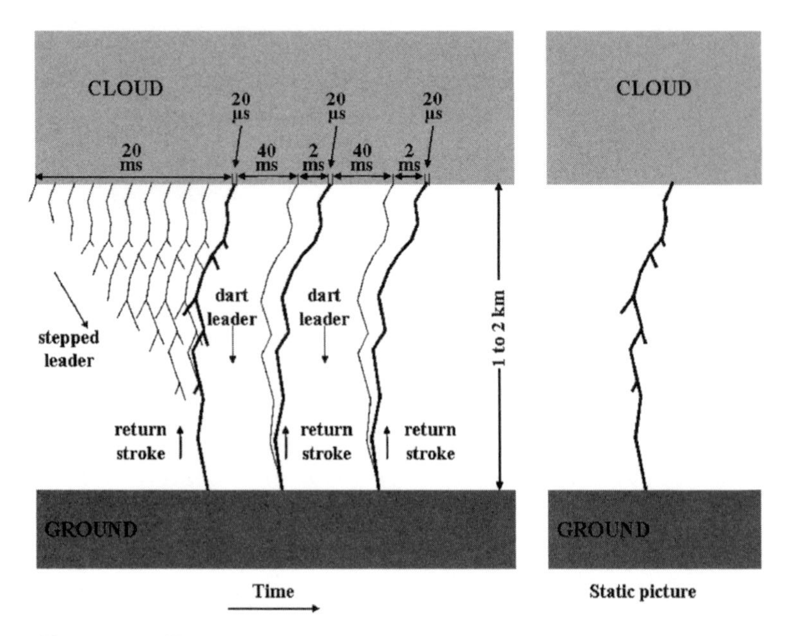

FIGURE 14. DEVELOPMENT OF THE DOWNWARD NEGATIVE FLASH IN THE CASE OF FLAT GROUND

ten meters in the case of flat ground (more in the case of tall structures). The return stroke or principal discharge propagates to the cloud at one third to one half the speed of light (this speed decreases with increasing height), neutralizing the leader charge partially or completely with an enhanced luminosity in the channel.

The rapidly increasing return stroke current heats up the leader channel up to a very high temperature (30,000 K) and generates very high pressure (some tens of atmospheres or more). This rapidly created hot plasma in ambient air is the generator of outward acoustic waves, i.e., thunder!

THUNDER!

Thunder is the acoustic emission of degenerated shock waves produced by the expansion of the rapidly heated lightning channel. The speed of sound in air is about 340 meters per second under standard conditions of temperature and pressure.

ELECTRIC CURRENT

An electric current, generated by a source, is the ordered flux of electric charges, which move, for example, in metallic conductors.

Let us consider a beam of positive electric charges q crossing a plane, which cuts it perpendicularly. The average current intensity i, during a time t, is the average value of the electric charge, q, crossing this plane during this time t, i.e.

$$i = \frac{q}{t}.$$

The unit of electric current is the ampere (A): 1 A = 1 C/s (coulomb per second). The kiloampere (1 kA = 1,000 A) is often used for convenience when discussing lightning.

Since lightning channels can extend for many tens of kilometers, the sound of thunder can last for many tens of seconds because it takes that long for the sound to reach us from the most distant parts of the discharge channel. Acoustic refraction perturbations, due to the geometric diversity of encountered obstacles and the decrease of temperature of the atmosphere with height can complicate the propagation of the sound of thunder. For these reasons, thunder can rarely be heard at distances greater than 25 km from the discharge channel. In the immediate vicinity of a lightning discharge, thunder comes as cracks, sometimes preceded by whistling.

The principal discharge (return stroke) of a negative CG flash is upward directed and develops a colossal peak electric power per meter of channel (hundreds of millions of watts per meter or 10^8 W/m) over a very short time interval, typically a few microseconds. The electric current reaches an average peak value of 30 kA in temperate regions. Some extreme values of the first-stroke peak current have been recorded on tall structures, reaching about 300 kA!

ELECTRIC POWER

Let q be a point charge moving in electric field between two points that are characterized by potential difference U.

The corresponding electric potential energy difference W = Uq varies at the following average rate:

$$P = \frac{W}{t} = U \frac{q}{t} = U\,i,$$

which is the electric power P.

The electric power is expressed in watts (W): 1 W = 1 J/s = 1 V·A (volt-ampere).

In general (in 80% to 85% of cases), the thundercloud is not completely discharged after a single stroke. A flash typically contains multiple (several) strokes instead of a single stroke. After an average pause lasting 60 ms (generally from a few to hundreds of ms), another leader, known as a dart leader, propagates continuously (not by steps), in the same channel, with a velocity of 1,000 to 10,000 km/s (*see figure 14*). This dart leader, not branched, follows, in opposite direction, the channel traversed by the first return stroke, generating a second return stroke. This stroke, a subsequent stroke, taps another electrically charged region of the thundercloud (*see figure 15*).

Typically, the dart-leader subsequent return-stroke sequences repeated three or four times. In one instance, 26 strokes were recorded in a single negative flash. A complete (multiple stroke) flash rarely lasts more than one second. The static picture that a human being can visually observe is shown in the right panel of figure 14.

The human eye cannot resolve the fast transient phenomena of a lightning flash. The retinal persistence being 0.1 second or so, our eyes are generally unable to resolve various components of a flash. Thus we use streak cameras and electronic devices.

Successive return strokes do not necessarily follow the same path as the first stroke, at least entirely. Note that 50% of the negative flashes show more than one strike point on the ground, individual strike points being separated by up to several kilometers.

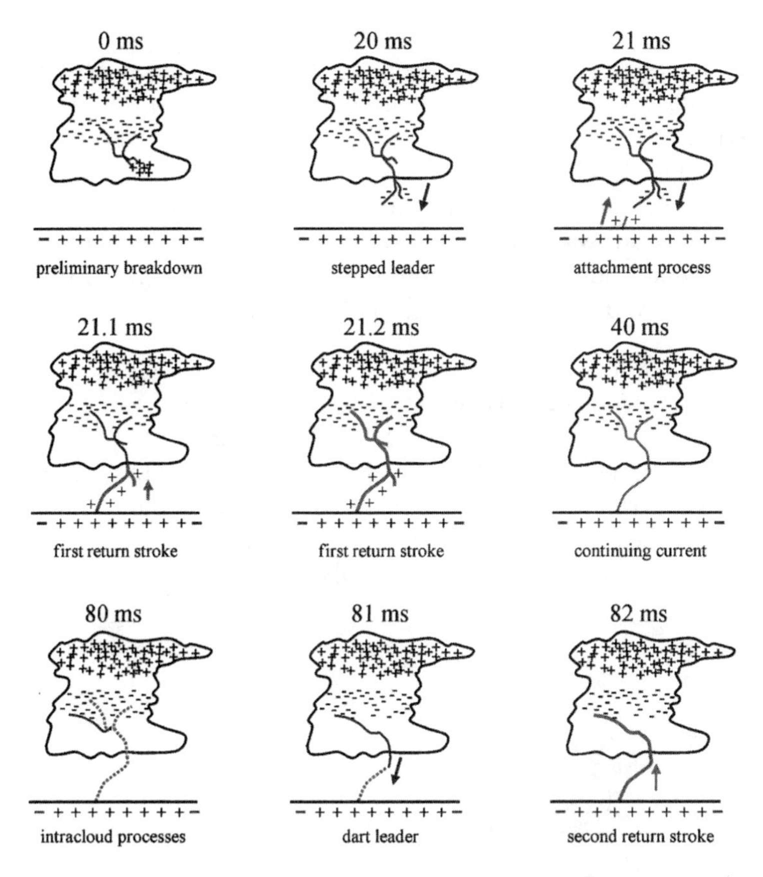

FIGURE 15. SOME IMPORTANT PHASES IN THE DEVELOPMENT OF A NEGATIVE CLOUD-TO-GROUND FLASH

Cloud-to-Ground Flash Phenomenology: Tall Towers

Every protrusion at ground level leads to an increased electric field at the local ground surface. For example, on a sharp ellipsoid with a ratio of large axis to small axis equal to 30 (*see figure 9, bottom right*), the electric field is enhanced by a factor of 300. As the air dielectric strength in uniform field is 30 kV/cm or 3 MV/m, an ambient field of 10 kV/m is enough to initiate a corona discharge at its tip. This is why luminous corona, or St. Elmo's fire, occurs above boat masts, on tall towers and church spires, on antennas, high-voltage line towers, treetops or even on the heads of hikers in the mountains. (St. Elmo, or Erasmus, was a martyr at Formia in Campania, under Diocletian. During thunderstorms, St. Elmo was invoked by seamen in the Mediterranean Sea.)

On tall structures (the taller they are, the larger the percentage of upward flashes) the subsequent strokes are similar to those in downward discharges to flat territory (*see figure 14*), but the initial leader develops from the structure (*see figure 16*), as opposed to developing from the *cumulonimbus*.

DOES LIGHTNING PREFERENTIALLY STRIKE TALL STRUCTURES?

Yes, it seems so. As the poet Jean de Sponde (1557–1595) already wrote in one of his sonnets: "The mountain is more often struck than the plain." On the other hand, lightning does not necessarily always strike the highest point (*see chapter 8*).

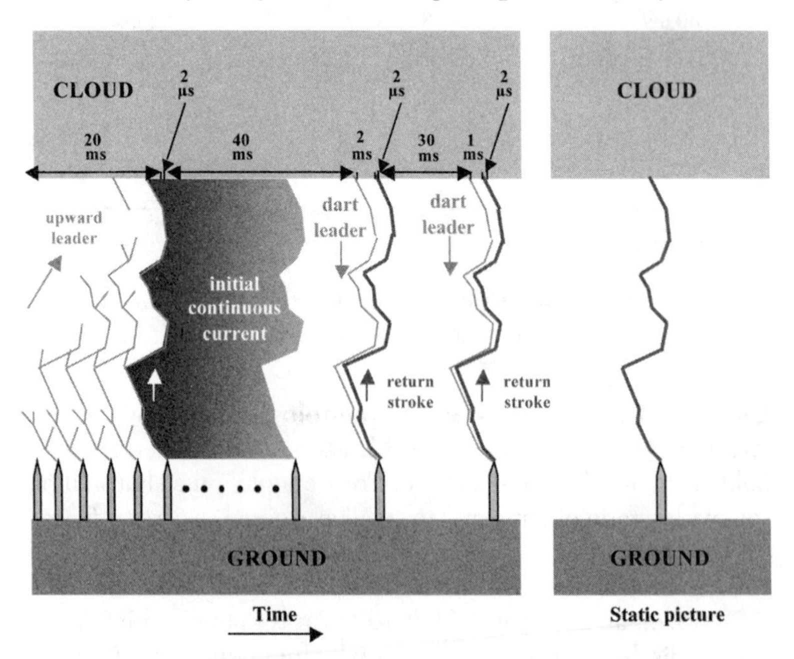

FIGURE 16. DEVELOPMENT OF AN UPWARD FLASH FROM A TALL STRUCTURE

Lightning Current Parameters

Figure 17 shows typical waveforms of negative lightning current (single stroke or the first component stroke of a multiple-stroke flash) and of positive lightning current (almost always single stroke). These currents are pulse-shaped waves with a variable duration, comprising a relatively sharp front (rising part from zero to the current peak value) and a longer tail (after the peak value). Some impulsive current components are followed by so-called continuing currents with relatively low magnitudes and durations up to hundreds of milliseconds.

The current wave of a single stroke or of the first component stroke of a multiple-stroke negative flash reaches its negative maximum value −1 p.u. (1 p.u. = 1 *per* unit = 100%) after a short time of a few microseconds, then slowly decreases to zero after some hundreds of microseconds.

The current wave of a positive flash, typically composed of a single stroke, reaches its positive maximum value of +1 p.u.

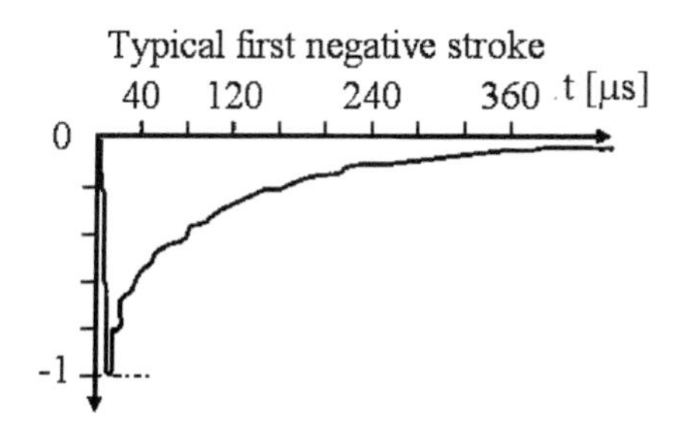

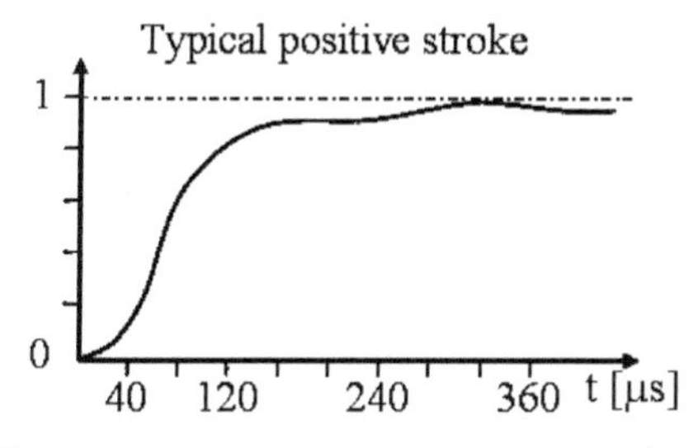

FIGURE 17. TYPICAL WAVEFORMS OF NEGATIVE (FIRST STROKE) AND POSITIVE LIGHTNING CURRENTS

in tens to hundreds of microseconds, then decreases to zero after some milliseconds (or more if continuing current is involved).

The Technical Committee 81 of the International Electrotechnic Commission (IEC TC 81: Lightning Protection) has adopted a simplified representation of the distribution of both positive and negative first strokes recorded all over the world. It consists of two straight lines shown in figure 18. The current amplitude I (in kiloamperes) is plotted on a horizontal axis. On the vertical axis is the probability P that a lightning current exceeds the value shown on the horizontal axis. For example, 98% of the recorded currents have a peak value larger than 3 kA and 1% of these currents have an amplitude larger than 200 kA.

Besides the current amplitude I (in kA), what are the other important parameters that can be derived from lightning current records? There are essentially seven useful parameters:

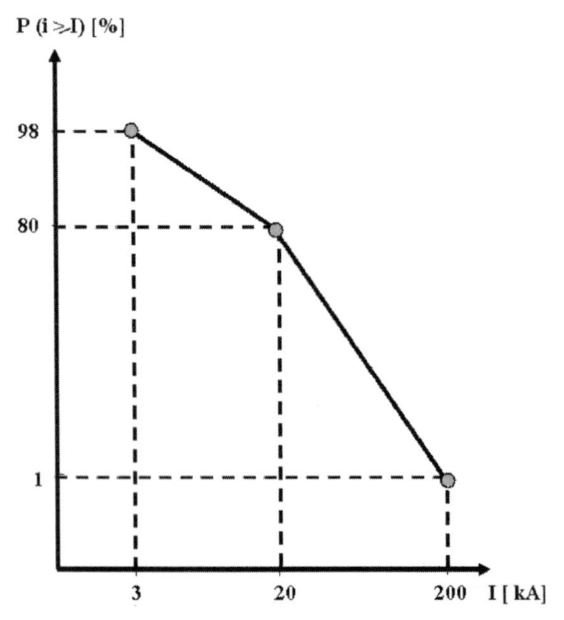

FIGURE 18. GRAPH OF THE LIGHTNING CURRENT PEAK VALUES (BOTH POSITIVE AND NEGATIVE STROKES)

(1) the electric charge (in C or coulombs); (2) the front duration of the current wave (in μs); (3) the maximum steepness of the current wave (in kA/μs); (4) each return stroke duration (in μs); (5) time intervals between successive return strokes of a negative flash (in ms); (6) the *total duration* of a lightning flash (in ms); and (7) the specific energy (or action integral) which is the energy dissipated by a lightning flash per unit electric resistance (in amperes squared times seconds or A^2s) .

The most important four parameters that are related to lightning effects are the current amplitude I, the maximum steepness of the current wave $(di/dt)_{max}$, the electric charge Q and the dissipated energy (Joule effect) in a 1-ohm resistor ($\int i^2dt$).

A positive correlation is found between the current amplitude I and the electric charge Q of a lightning flash. There is another positive correlation between the current amplitude I and the specific energy $\int i^2dt$ when we neglect contributions from continuing currents. If we only consider the first stroke of a negative multiple-stroke flash, a strong correlation appears between the current amplitude I and the electric charge Q and the specific energy $\int i^2dt$. Positive correlation is also found between the total electric charge Q of the flash and its duration t. For the other parameters, correlations are much less significant.

The table in figure 19, based on data of Karl Berger and adopted by CIGRE (*Conseil International des Grands Réseaux Électriques à Haute Tension* or International Council of Large High-Voltage Electric Power Systems), contains the 95%, 50%, and 5% values of the different parameters derived from measured lightning currents.

These data are from Switzerland. There exist similar data from other countries.

For example, in the third column, only for positive flashes, we have a 5% probability to get a current amplitude larger than 250 kA, an electric charge larger than 350 C, a front duration larger than 200 μs, a maximum steepness larger than 32 kA/μs, a pulse duration larger than 2 ms, and a specific energy larger than 15×10^6 $A^2.s$.

Parameters	Unit	Percentage exceeding tabulated value		
		95%	50%	5%
Peak currents				
first negative strokes and negative flashes	kA	14	30	80
subsequent negative strokes	kA	4.6	12	30
positive flashes	kA	4.6	35	250
Charge				
first negative strokes and	C	1.1	5,2	24
negative flashes	C	0.2	1.4	11
subsequent negative strokes	C	1.3	7.5	40
positive flashes	C	20	80	350
Front duration				
first negative strokes	μs	1.8	5,5	18
subsequent negative strokes	μs	0.22	1.1	4.5
positive flashes	μs	3.5	22	200
Maximum rate of rise (di/dt)				
first negative strokes	kA/μs	5.5	12	32
subsequent negative strokes	kA/μs	12	40	120
positive flashes	kA/μs	0.2	2.4	32
Pulse duration				
first negative strokes	μs	30	75	200
subsequent negative strokes	μs	6.5	32	140
positive flashes	μs	25	230	2000
Time intervals between				
negative strokes	ms	7	33	150
Flash duration				
negative (simple or multiple)	ms	0.15	13	1100
negative (multiple only)	ms	31	180	900
positive	ms	14	85	500
i^2dt integral				
first negative strokes and negative flashes	$A^2.s$	6.0×10^3	5.5×10^4	5.5×10^5
subsequent negative strokes	$A^2.s$	5.5×10^2	6.0×10^3	5.2×10^4
positive flashes	$A^2.s$	2.5×10^5	6.5×10^5	1.5×10^7

FIGURE 19. PARAMETERS DERIVED FROM MEASURED LIGHTNING CURRENTS

CAN WE UTILIZE ENERGY FROM LIGHTNING?

Even though the power associated with each lightning stroke at a point of impact is tremendous, the short duration (1 ms or so) makes the associated energy rather low. Furthermore, capturing lightning strikes would require a large number of tall towers, which is impractical. It would not be practical to invest in such a natural source of energy!

In summer, 90% of the flashes are negative, only 10% are positive. In winter, it is different: the majority of flashes can be positive. Positive flashes can also be dominant in the dissipating stage of thunderstorms and in some severe storms.

ELECTRIC RESISTANCE

Any electric current experiences a certain resistance from the medium in which it flows. This resistance necessarily leads to an energy loss. In 1826, Georg Simon Ohm (1789–1854) proved that the resistance R to the electric current flow in a conductor depends on its geometric shape, its material and the temperature.

Ohm showed that, for a given conductor and a fixed temperature, a potential difference U across the conductor is proportional to the current I in the conductor. The proportionality factor is called (electric) resistance R of the conductor:

$$U = R\, i.$$

This is Ohm's law, which is of great practical value, although it applies only to ohmic conductors (usual metals and some non-metallic conductors). The unit of (electric) resistance is the ohm (symbol Ω): 1 Ω = 1 V/A.

The power dissipated in a resistor R is:

$$P = R\, i^2 \quad \text{or} \quad P = \frac{U^2}{R}.$$

James Prescott Joule (1818–1889) was the first scientist to verify experimentally, in 1841, that the heating power of an electric current in a resistor (Joule effect) is of this kind.

A negative flash to ground can last from less than a millisecond ("cold" flash) to more than one second ("hot" flash). A flash is composed of a single stroke in 15–20% of the cases. A flash may have multiple strokes in 80–85% of the cases, an average flash comprising from 3 to 5 components (strokes). Peak current values of a first negative component (average value around 30 kA) are generally 2 to 3 times larger than those of subsequent components (average value around 10–15 kA). However, about one third of the flashes to ground contain a subsequent component with a peak current value that is larger than the first component. The maximum steepness of the current wave is typically 100 kA/μs or so.

Leader/return-stroke sequences constitute the first mode of electric charge transfer to ground. Some return strokes are followed by continuing currents, which represent the second mode of electric charge transfer to ground. This quasi-steady current has amplitude of about 100 A. It flows through the plasma channel during one tenth of a second or so and transfers the resulting electric charge of some tens of coulombs.

A third mode of electric charge transfer to ground was identified by Rakov and co-workers. Optical manifestations of this mode were observed earlier by Malan and Collens (in 1937) as temporary enhancements in luminosity of the faintly luminous continuing current channel, called M-components. The corresponding current pulses have rise times in the range of 300–500 μs, and current amplitudes between 100 and 1,000 A, dissipating electric charges of 0.1–0.2 C (*see figure 20*). According to V. Rakov (*see references*), M-components result

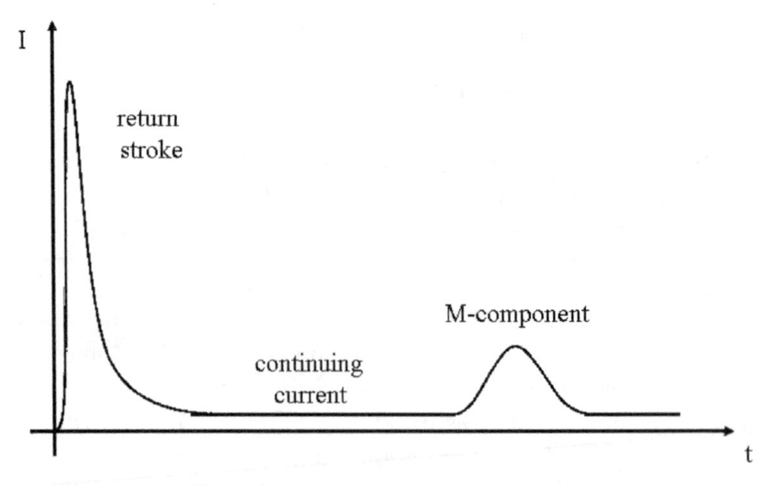

FIGURE 20. RETURN STROKE FOLLOWED BY A CONTINUING CURRENT AND AN M-COMPONENT

from the superposition of two guided current waves progressing in opposite directions: the first one is progressing downward as a leader and the second one upward, the latter one being reflected from the ground which behaves essentially as a short-circuit.

Positive flashes are generally composed of a single stroke, but usually contain large continuing currents that transfer a much larger positive charge to ground than negative flashes.

Bipolar discharges to ground constitute a new field of study for researchers. People interested should refer to Rakov's and Uman's book (*see references*).

Transient luminous events, some related to lightning, also occur in the middle atmosphere above cloud tops and below the ionosphere. These include blue jets, red sprites, and elves. Recently it has been shown that thunderstorms emit X-rays and so-called terrestrial gamma flashes.

Red sprites are luminous events developing in the mesosphere (at altitudes of 45–85 km). Most often, they are associated with positive discharges to ground (*see color insert*). Further research is needed in this field (*see chapter 10*), since we presently know too little about positive discharges.

The elves look like glowing pancakes, generated between 75 and 100 km above ground by relatively strong lightning flashes. They are initiated by electromagnetic pulses produced by lightning return strokes. These impulsive fields accelerate electrons that hit and excite air molecules. By de-excitation, these molecules generate some visible light. This phenomenon creates luminous expanding rings.

Blue jets, about 40 km high, are infrequently observed. They extend upward at a high velocity (~120 km/h) from the cloud top, but the formation mechanism is still unknown. They are apparently not related to ordinary lightning.

X-rays produced by thunderclouds are possibly related to lightning initiation. The terrestrial gamma flashes, recorded by satellites, are apparently related to thunderstorm activity and are the first manifestation of gamma ray creation in the terrestrial atmosphere, besides nuclear reactions . . . a new puzzle to be solved!

RESISTIVITY AND ELECTRICAL CONDUCTIVITY

Ohm showed that the resistance of a conductive wire is proportional to its length l and inversely proportional to its cross-section S. The proportionality coefficient is resistivity ρ characterizing the material, so that

$$R = \rho \frac{l}{S}.$$

This relation is sometimes called Pouillet's law. At ambient temperature, for example, copper has a resistivity ρ equal to $1.7\ 10^{-8}\ \Omega m$. Materials with a resistivity lower than $10^{-5}\ \Omega m$ are (electrical) conductors, materials with a resistivity higher than $10^{+6}\ \Omega m$ are insulators (dielectrics) and materials with resistivity between these two limits are essentially semiconductors. Resistivity depends on temperature T. Experiments show that the resistivity ρ of most materials varies almost linearly with small variations of temperature ΔT.

The (electrical) conductivity σ is defined as the inverse of the resistivity ρ,

$$\sigma = \frac{1}{\rho}.$$

It is measured in siemens per meter (S/m). $1\ S = 1\ \Omega^{-1}$ (1/ohm).

The Global Electric Circuit

Due to ionization by cosmic rays and Earth's natural radioactivity, the electrical conductivity of air is not zero near the ground surface and quickly increases with altitude. The ionosphere acts as a conductor. The Earth's surface is negatively charged (a 500,000 C negative electric charge is maintained continuously on the Earth's surface) and the ionosphere can be viewed as being positively charged. The vertical downward component of the electric field at the surface is about 100 V/m in fair weather conditions.

As a result, we have a huge capacitor (*see figure 21*) made of two almost spherical electrodes: the ionosphere (electrosphere)

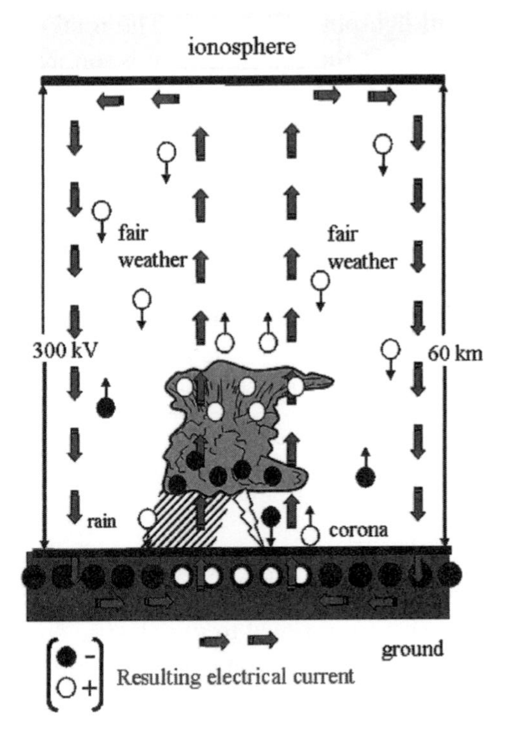

FIGURE 21. THE GLOBAL ELECTRIC CIRCUIT: ATMOSPHERIC CAPACITOR FORMED BY THE IONOSPHERE (ELECTROSPHERE) AND GROUND THAT IS RECHARGED BY THUNDERSTORMS

and the ground surface. In fair weather, the resistance of a vertical air column with a 1 cm^2 cross-section and a length of 60 km is very high and the global resistance for the whole Earth's surface is 300 ohms. Applying Ohm's Law, a total leakage current of 1,000 amperes or so leads to a voltage of 300 kV between the electrodes (ionosphere and ground). The re-charging of this huge, leaky capacitor comes from the thunderstorm activity (on the whole Earth about 100 flashes occur every second).

Without thunderstorms, the atmospheric capacitor would discharge completely on a time scale of roughly ten minutes.

The fair-weather current, of the order of 1 kA, in the global electric circuit must be balanced by the total generator current, which is composed of currents associated with corona, precipitation, and lightning discharges. The total current flowing from cloud tops to the electrosphere is, on average, about 0.5 A per thunderstorm. Negative charge is brought to Earth mainly by cloud-to-ground lightning discharges (most of which

CAPACITANCE OF A CAPACITOR AND ELECTRIC POTENTIAL ENERGY

A capacitor is a device used to store electric field energy. At a fixed potential difference U, the charge Q on each of the two capacitor plates depends on its geometry and characteristics of the dielectric between the plates, that is, on electric capacitance:

$$Q = C\,U.$$

The larger the capacitance, the greater the charge stored at a given potential difference (voltage). Thus, the capacitance C is directly proportional to the electric charge Q and inversely proportional to the voltage U. Its unit is the farad (F): 1 F = 1 C/V.

In practice, capacitances encountered are generally expressed in μF (1 microfarad is one millionth of one farad), in nF (1 nanofarad is one billionth of one farad) or in pF (1 picofarad is one trillionth of one farad).

Here are some values of capacitances met in practice: 1 pF for a needle tip, 150 pF for a human body, 500 pF for a car, 1 nF for a big truck.

When applying a voltage U to the capacitor plates, a total charge +Q is placed on the positive plate and a total charge −Q on the negative plate, though initially we had Q = 0 and U = 0. To transfer the charge, a certain work must be done; it can be expressed in three different ways:

$$W = \frac{1}{2}\frac{Q^2}{C} = \frac{1}{2}C\,U^2 = \frac{1}{2}Q\,U.$$

This work corresponds to the electric potential energy (energy stored in the capacitor of capacitance C).

transport negative charge to ground) and by corona current under thunderclouds. The net precipitation current is thought to transport positive charge to ground. Since there has been no experiment to confirm conclusively this classical picture of global electric circuit, it remains a subject of debate.

References

V. Cooray, *The Lightning Flash*, IEE, Power & Energy Series 34, UK, 2003.

V.A. Rakov and M.A. Uman, *Lightning – Physics and Effects*, Cambridge Univ. Press, 2003.

5
Where Does Lightning Strike?

Lightning Incidence, Detection, and Mapping

Surface heating, a steep decrease in temperature with height, and abundant moisture, are generally necessary for the formation of thunderstorms. If any one of these ingredients is missing (for example, in deserts or polar regions), thunderstorms rarely occur. In tropical regions, thunderstorms are frequent. In Bogor, in the western part of Java Island, thunder is heard almost every other day. In western Europe, thunderstorms occur only ten to thirty days a year.

The number of thunderstorm days per year (unit: $year^{-1}$) T_d or the keraunic level is the average number of days per year when thunder can be heard. Though used in the past, it is actually not a very useful parameter since, in temperate regions, a thunderstorm can last for tens of minutes or several hours, depending on the meteorological conditions. Sometimes thunder can be heard at unusually large distances, say, 40 km or even more, giving a strongly exaggerated impression of the lightning activity, although most of the time thunder cannot be heard more than about 25 km from the lightning that causes it. Whether thunder can be heard can also depend strongly on local conditions. Loud sounds, such as, for example, the sounds of jet aircraft at an airport, may mask distant thunder.

There are some scientists who have suggested that the frequency of occurrence of thunderstorms on Earth could be related to the solar activity with a periodicity of 11 years (solar cycle), though the connection is still subject to debate.

In France, by means of the Météorage network of electric and magnetic field antennas, employing triangulation in real time, it has been shown that the regions which were most often struck by lightning were the Southern Alps, the Pyrenees (especially Western Pyrenees) and the Massif Central where the number of thunderstorm days per year is greater than thirty. On the Northwest Coast, along the Channel, it is weaker, between ten and eighteen. For the whole of France, an average of twenty thunderstorm days per year is generally assumed.

In Belgium, the Royal Meteorological Institute has installed a SAFIR lightning detection system. It employs the method of electromagnetic interferometry and allows one to track thunderstorms in real time in the whole country (*see color insert*). An average value of fifteen thunder days per year is generally accepted (average of values between eight and twenty-two depending on the regions) for Belgium.

In the intertropical zone, in Central South America (from Colombia and Peru to the Center-South part of Brazil), in Central Africa (from Guinea to Tanzania and South Africa) and in Indonesia, the keraunic level can be larger than 100 per year.

Figure 22 shows the 1956 world map of *isokeraunics* (curves of same value of the number of thunder days per year) compiled by the World Meteorological Organization (Geneva, Switzerland, publication # 21).

The keraunic level is an indicator of thunderstorm activity. It is not rigorous at all since it gives no indication of the number of lightning strikes to ground. It has been supplanted by the ground flash density N_g, i.e., the number of lightning flashes to ground per kilometer squared per year (unit: $km^{-2}.year^{-1}$). In temperate regions, N_g in $km^{-2}.year^{-1}$ is roughly one tenth of the keraunic level T_d ($year^{-1}$).

In France, this corresponds to $1\ km^{-2}.year^{-1} < N_g < 3.5\ km^{-2}.year^{-1}$ with an average value equal to $2\ km^{-2}.year^{-1}$. In Belgium, it is $0.8\ km^{-2}.year^{-1} < N_g < 2.2\ km^{-2}.year^{-1}$ with an average value equal to $1.2\ km^{-2}.year^{-1}$. In Brazil, Indonesia, Florida, and in Central Africa, N_g is much larger, generally between 8 and $15\ km^{-2}.year^{-1}$.

FIGURE 22. WORLD MAP OF ISOKERAUNICS (WORLD METEOROLOGICAL ORGANIZATION)

The U.S. National Aeronautics and Space Administration (NASA) has placed optical detectors on satellites to record the transient luminous signals emitted by all types of lightning flashes (IC and CG) on Earth without discrimination between types.

There are many factors influencing lightning incidence. The following parameters are important to consider: topographical factors (soil humidity, thunderstorm corridors favored by airstreams in valleys, lightning strikes on hillsides instead of mountaintops, etc.) and geological and orohydrographical factors (faults, crevices, cracks, water layers, etc.). These and other factors can be responsible for the observed inhomogeneity of spatial distribution of lightning ground flash density.

IS IT TRUE THAT LIGHTNING NEVER STRIKES THE SAME PLACE TWICE?

This is not true. Consider sky-scrapers: they are sometimes struck several times during the same thunderstorm. For example, Ostankino Tower in Moscow is struck (mostly by upward flashes) on average thirty times per year.

Can We Artificially Initiate Lightning?

The first triggered lightning discharges were initiated in 1960 from a research vessel close to the west coast of Florida. Small rockets were launched upward to thunderclouds, unwinding thin-grounded wires behind them.

In 1973, French researchers developed at Saint-Privat-d'Allier (Massif Central in France), the first French triggering station, using anti-hail rockets equipped with a spool of Kevlar-coated copper wire, one end of which was connected to ground (*see figure* 23), allowing the triggering of lightning. A Belgian team, under Bouquegneau's assistant Pierre Depasse, collected data showing that this triggered lightning had many

FIGURE 23. LAUNCHING OF A LIGHTNING-TRIGGERING ROCKET AT SAINT-PRIVAT-D'ALLIER (FRANCE)

characteristics similar to natural lightning, except for the first return stroke of the negative flash.

The lightning-triggering facility at Camp Blanding, Florida was established in 1993 by the Electric Power Research Institute (EPRI) and Power Technologies, Inc. (PTI). Since September 1994, the facility has been operated by the University of Florida (UF). More than forty researchers (excluding UF faculty, students, and staff) from fifteen countries representing four continents have performed experiments at Camp Blanding concerned with various aspects of atmospheric electricity, lightning, and lightning protection. Since 1995, the Camp Blanding facility has been referred to as the International Center for Lightning Research and Testing (ICLRT). Over a fourteen-year period (there was no triggering in 2006), the total number of triggered flashes has been 317, that is, on average about twenty-three per year, with sixteen (about 70%) of them containing return strokes. Out of the total of 317 flashes, 314 transported negative charge and three either positive or both negative and positive charge to ground.

The principal results obtained at the ICLRT include characterization of the close lightning electromagnetic environment, identification of the M-component mode of charge transfer to ground, inferences about the interaction of lightning with ground and with grounding electrodes, discovery of X-rays produced by triggered-lightning strokes, first direct measurements of NOx production by lightning, direct estimates of lightning input energy, and ground-truth evaluation of the performance characteristics of the U.S. National Lightning Detection Network.

The "rocket-and-wire" method used to trigger lightning is presently the only operational one. New techniques, including laser triggering, triggering by microwave beams, triggering by water jets, and triggering by transient flames have been proposed. None of these new techniques have been unambiguously demonstrated to be capable of initiating lightning so far.

Does Lightning Exist on Other Planets?

Charge separation in a variety of non-thunderstorm situations, from sandstorms to volcano eruptions, earthquakes, and nuclear explosions, can result in the production of lightning-like electrical discharges.

What are the necessary conditions on other planets to facilitate the occurrence of lightning or lightning-like discharges? These planets have to contain particles of at least two different types or of the same type but with different properties (dimensions, temperatures, etc.) whose interaction results in local charging so that charges of opposite sign are acquired by the different classes of particles. Moreover, differently charged particles must be separated by distances (kilometers) comparable to the ones involved in terrestrial discharges.

Lightning or lightning-like discharges surely exist on Jupiter and Saturn (*see recent discoveries by Cassini spacecraft*). A very strong thunderstorm was recorded on Jupiter, on December 7, 2000 (*see figure 24*). The famous red spot on Jupiter, twice as big as our Earth, was already observed by Robert Hooke in 1664 and constitutes a permanent gigantic whirlwind at high pressure.

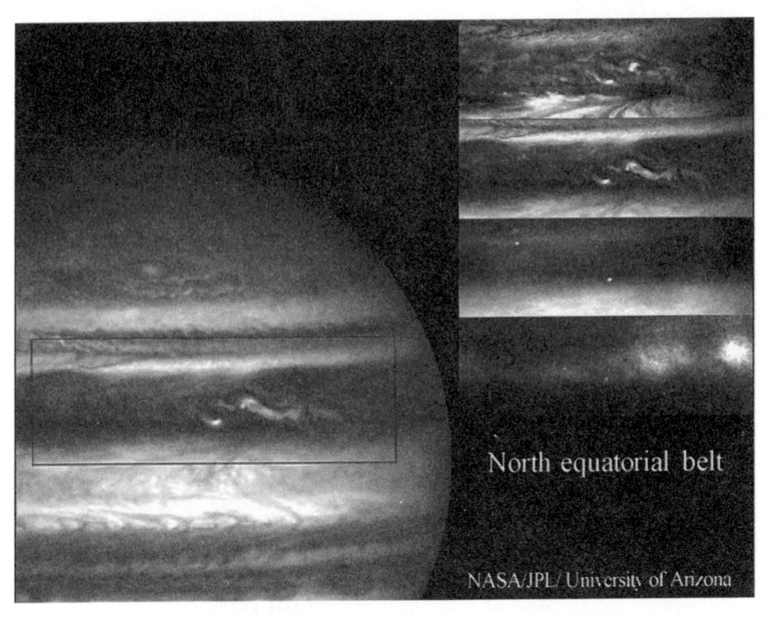

FIGURE 24. THUNDERSTORM ON JUPITER (CASSINI SPACECRAFT, DECEMBER 7, 2000 NASA DOCUMENT)

Jupiter and Saturn are the only two planets from which optical signals attributable to lightning have been observed.

Lightning may exist on Venus, but if it does its phenomenology should be completely different from that of Earth. While there is considerable evidence in the surface topography of previous volcanic activity, the Venusian atmosphere below about 30 km is clear and shows no ejected volcanic material.

Mercury and Mars have tenuous atmospheres without volcanic activity. They should not experience lightning, although severe dust storms can initiate electrostatic discharges on Mars.

Lightning could also exist on Io (Jupiter's satellite) and Titan (Saturn's satellite), maybe on Uranus and Neptune (and their satellites) although nothing unambiguous (for example, light signals) has been observed so far. We presently know almost nothing about the couple Pluto-Charron.

References

P. Depasse, *Statistics on artificially triggered lightning*, J. Geoph. Res. 99, D9, 1994.

V.A. Rakov and M.A. Uman, *Lightning – Physics and Effects*, Cambridge Univ. Press, 2003.

Part III
Lightning Effects

Like any electric current, the lightning current obeys the usual laws of electromagnetism. However, the strong electric fields of lightning make it possible for it to form a conducting channel through weakly conductive or non-conductive media. To minimize damage to any object that might be struck by lightning, this current must be offered a conductive path that is as direct as possible. This path must be connected to all neighboring metallic elements and to ground.

One of the best descriptions of modern lightning protection principles was given by P.G. Laurent in 1950:

The lighting current is an electric current like other currents. It propagates following ordinary laws of Electrotechnics. You can foresee its behaviour as long as the system configuration can be specified and the current flow can be simulated with pulse generators at lower voltages. We think that all practical means of lightning protection boil down to only one: to offer the current a conductive path as direct as possible and to interconnect all neighboring metallic elements to it. Lightning (femininely personalized in French and in Russian!) *is an important figure to whom you do not easily resist without danger, but who accepts to be directed rather easily when you submit to her desires. She has millions of volts available to break down insulating obstacles, but flows unnoticed in relatively small diameter conductors. When progressing downward, she has the curiosity to explore grounds or neighboring metallic conductors where she expects, right or wrong, to find an easier outlet, it is better to help her with convenient interconnections than to oppose her with obstacles which generally run the risk to turn out to be a weak point.*

This somewhat animist language translates to the physical reality: lightning "sees" the neighboring space thanks to some kind of sensitivity which is offered by the electric field that she carries with her, and this field gives her a means of action, including piercing insulators when the electric field exceeds its critical breakdown value.

The essential idea of any protection system must be to deprive her opportunities to put these means at work.

Lightning effects result from the presence of current in a lightning channel and in conductors (*see chapter* 6). These effects are electric (charges, currents), electrodynamic (forces), thermal (heat emission), electromagnetic (radiated fields inducing dangerous over-voltages in electric circuits), electrochemical (galvanic decomposition), acoustic (thunder, pressure shock waves) and physiological (effects on heart and the nerves, those that control respiration in particular).

A distinction is made between direct effects (direct striking of structures leading to mechanical damages, fires, melting, etc.) and indirect effects (striking ground or structures at a distance from the structure in question, generating over-voltages which propagate, by conduction or radiation, to devices inside the structure to be protected, *see chapter* 7). Against direct effects, protection generally consists on installing lightning rods or, more precisely, an external (or structural) lightning protection system (external LPS). Against indirect effects, protection generally consists of installing coordinated surge protective devices (SPDs) associated with various electrical and electronic circuits.

6
Physical Effects

Electrical Effects

When lightning current is injected into ground, which usually has some resistance, the electric potential at the strike point can rise to a very high value. This leads to local soil breakdown, which can damage buried electrical conductors. This potential rise is a serious hazard for people on the ground surface (step voltage) in the vicinity of the strike point. Moreover, electrical effects due to galvanic coupling or conduction coupling create surges on transmission lines and telecommunication lines, as well as on the earth termination systems.

Electrodynamical Effects

Between parallel (or quasi parallel) conductors carrying currents in the same direction (antennas, down conductors, etc.) attractive forces can lead to mechanical damages (these conductors can bang together or even crush).

These effects are typically negligible if the conductors are separated by more than about 50 centimeters. Indeed, for a current of 100 kA, attractive forces reach 400,000 newtons per meter, i.e., the equivalent of an enormous linear force of 40 tons per meter for 5 mm spacing, but only 400 newtons per meter for 50 cm spacing.

These electrodynamic effects are different from the blast effects (*see acoustic effects*) that are able to break glass and walls, to cause deafness and serious internal hemorrhages, as well as explosive bursting of tree bark (*see figure* 25) and shat-

FIGURE 25. TREE STRUCK BY LIGHTNING: BURST BARK

tering of non-conductive elements of structures, particularly when moisture is present.

Acoustic Effects

An acoustic shock wave is created in the immediate vicinity (within a few meters) of a lightning channel (*see preceding section*) due to the rapid increase in temperature and consequent explosive expansion of the channel. The initial propagation speed of the shock wave is about ten times the speed of sound (the latter being about 340 meters per second). This speed decreases rapidly, and within a few meters the shock wave is transformed into an acoustic wave that propagates at the speed

1. Multiple Strokes (Gullegem, Belgium).

Photo by C. Bouquegneau

2. Stone Engraving of a Sorcerer Lightning God (4,000 years ago), Vallée des Merveilles (South of France)

Photo by C. Bouquegneau

3. Thunderbird on a Totem Mast (Vancouver Island, Canada, 20th Century)

Photo by C. Bouquegneau

4. Raiden (Raijin), Japanese God of Thunder (Temple Sanjü-Sangendö, Kyoto, 13th Century)

Photo by C. Bouquegneau

5. Namarrgon, The Lightning Man (Kakadu National Park, Northern Australia) ; detail on the right

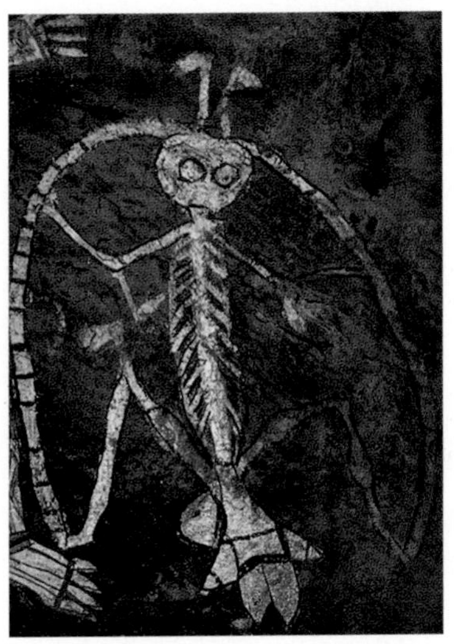

Photo by C. Bouquegneau

Photo by C. Bouquegneau

6. Dordje (Diamond Lightning, Vajra) and Drîlbu (Bell, Ghanta) Nepalo-Tibetan (15th Century)

7. Cloud-to-Ground, Cloud-to-Cloud and Intra-Cloud Flashes (Duinbergen, Belgium)

8. Positive Cloud-to-Ground Flash (Duinbergen, Belgium)

9. Different Types of Lightning Flashes (Duinbergen, Belgium)

10. Cloud-to-Air Discharge (Duinbergen, Belgium)

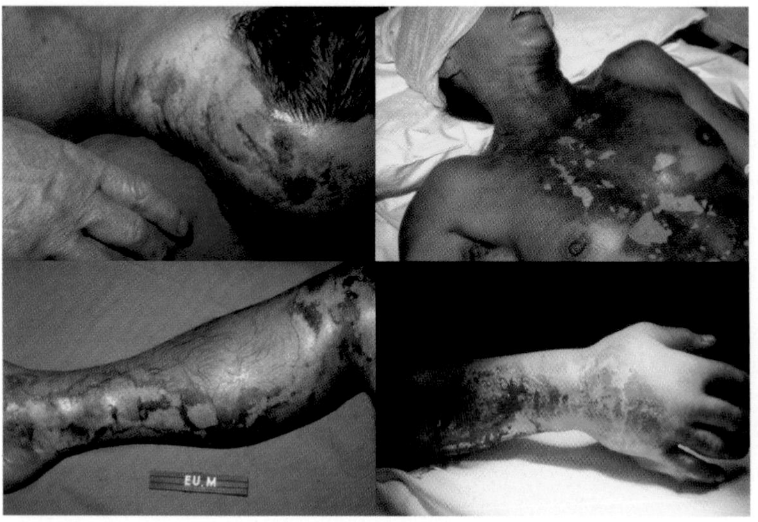

11. Various Superficial Burns Due to Lightning

12. Cows Killed by Lightning Under a Tree (Step Voltages)

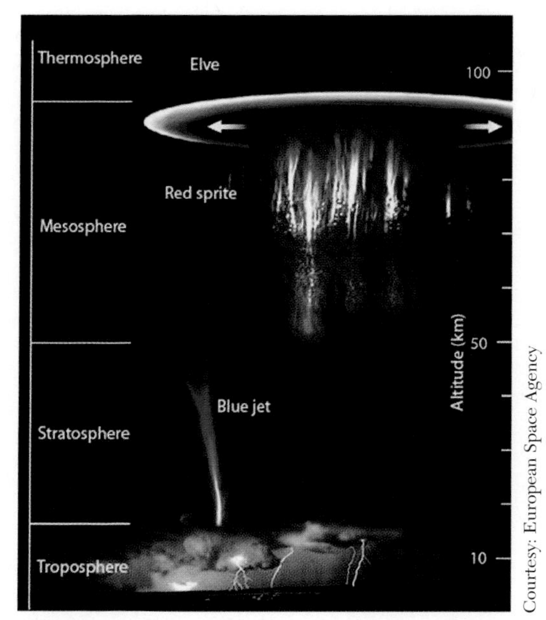

Courtesy: European Space Agency

13. Elves, Sprites, Blue Jets:
Stratospheric and Mesospheric Lightning

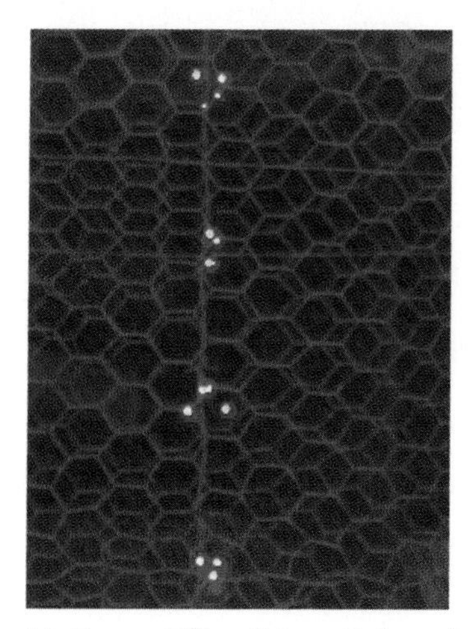

14. Corona Effect (Mons, Belgium)
High-Voltage Laboratory, 2006

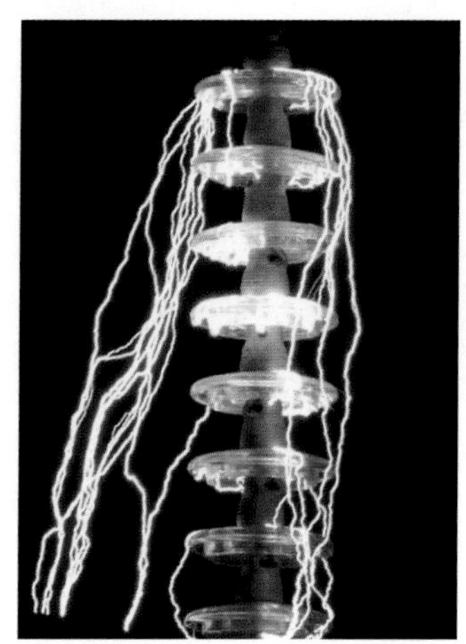

15. High-Voltage Laboratory
Discharge in Alternating Current
(Not a Lightning Flash!)

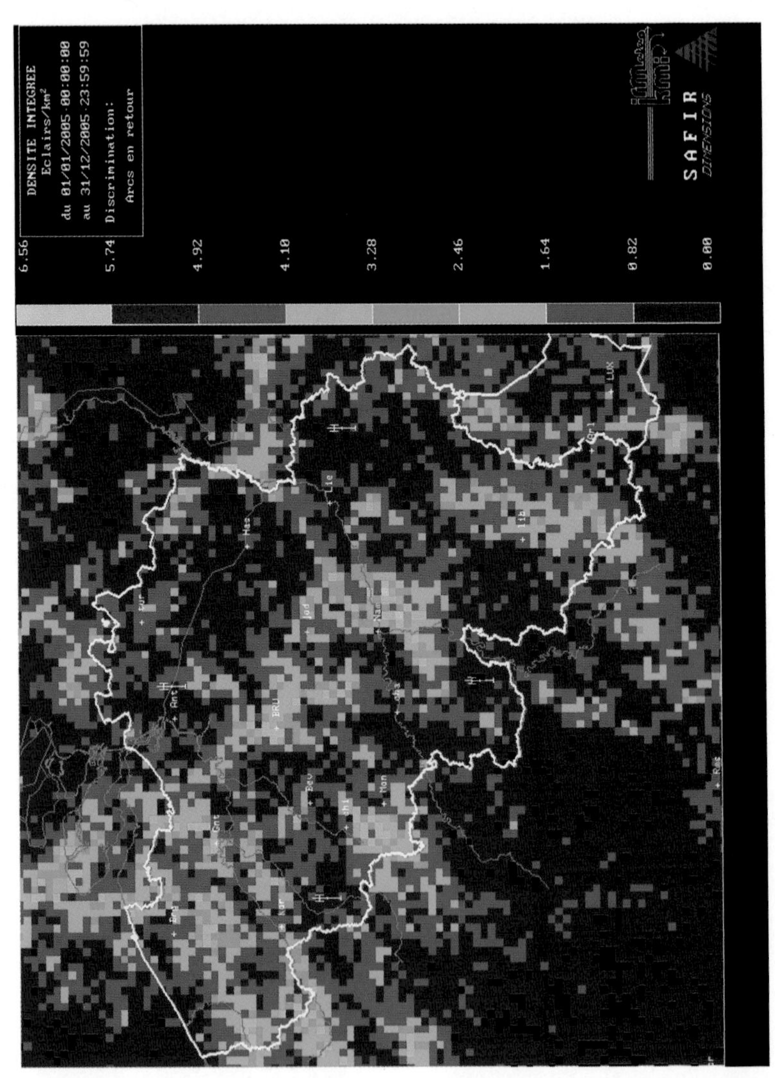

16. SAFIR Map of the Lightning Flash Density in Belgium for the Year 2005, Royal Meteorological Institute (Brussels, Belgium)

of sound. Thunder can be viewed as a degenerated shock wave. Thunder can seldom be heard beyond about 25 km or so.

Thermal Effects

A lightning discharge can lead to the fusion of metallic elements at the strike point. "Hot" flashes (containing long continuing currents) can ignite dry wood and provoke fires. Sandy soils (silica) can be vitrified by the current flow (plasma at 30,000 K peak temperature) and fulgurites can be formed (*see figure 26*). Fulgurites follow the shape of the lightning channel in soil; they are sometimes called fossilized flashes. They can reach a

FIGURE 26. A PIECE (4.6 M LONG) OF A FULGURITE SAMPLE (35 M LONG) EXCAVATED IN THE LIBYAN SAHARA DESERT BY THE FRENCH MINERALOGIST A. LACROIX (MINERALOGY MUSEUM IN DRESDEN, GERMANY)

length of several tens of meters (a 35-meter-long fulgurite was excavated in the Libyan desert).

On flat metallic surfaces, if we suppose that the anode potential drop U at the base of the electric arc remains constant during the whole discharge, the locally dissipated energy is W = Q U (*see potential energy discussion in chapter 4*). For example: a typical flash (Q = 30 C, U = 20 V, W = 600 J) causes the fusion of approximately 60 mm^3 of steel, corresponding to a diameter of 25 mm and a penetration depth of 0.15 to 0.25 mm (0.4 to 0.6 mm in copper or aluminum), assuming that no heat is dissipated in the metal mass. Even a much bigger than average flash (Q = 300 C) can only perforate steel sheets less than 2 or 3 mm thick. This is why lightning flashes only fuse rather thin metallic wires, except at the strike point.

Long-lasting discharges with continuing currents are generally thought to be responsible for setting fires. Additional heating can be produced due to poor contacts. Moisture inside wood or masonry (particularly in cracks, joint, gaps, also tree sap) can be turned to steam and the resulting thousand-fold increase in pressure leads to an explosion.

Electromagnetic Effects

Electromagnetic effects include potential differences induced in open loops. These potential differences can produce sparks igniting nearby flammable or explosive materials. Electromagnetic effects can cause over-voltages in relatively low-voltage distribution or telecommunication lines when the lightning strike point is within some hundreds of meters of the line. These over-voltages can cause flashover and line outage or propagate to electric and electronic equipment at line terminations causing damage there in the absence of adequate protection.

Electrochemical Effects

Even on lightning down conductors that frequently carry lightning currents, electrochemical decomposition does not occur. Corrosion is accelerated by earthing currents on buried conductors (e.g., cables, pipes, earth-termination systems). Nevertheless, since lightning currents are of relatively short duration, these effects are generally very low with respect to those generated by telluric currents (electric currents circulating inside the earth), particularly in the superficial layer possibly involving metallic elements at ground level. These currents are due to electric fields induced by the Earth magnetic field variations in the slightly conductive ground.

Lightning Effects on Humans and Animals

Physiological effects range from merely being stunned to almost instantaneous death, neurological effects, loss of vision, cataracts, deafness or ruptured eardrum, paralysis, temporary fainting (sometimes with short respiratory arrest), short or long-duration comas.

If lightning current flows through the body it can lead to serious or even fatal injuries. However, sweaty skin and wet clothes can offer a preferred path for the electrical discharge. Clothes can blow apart from the pressure wave and shoes can be blown off. This thermal shock is so short-lived that only superficial burns can occur, but metallic objects (i.e., necklaces) can reach high temperature (at least superficially) leading to deeper burns.

Generally burns are superficial (deeply cutaneous close to the incoming and outgoing points, linearly superficial corresponding to the quick bypassing electrical discharge, superficial but spread out by the electrical arc). Those occurring because of hot metallic objects can be more serious.

Lightning victims can also have erythemateous treelike discharges or Lichtenberg figures which are keraunographic fractally shaped prints (*see figure 27*, after C.W. Bartholomé: *Cutaneous manifestations of lightning injury*, Arch. Dermatology, 111 (11) 1466-8/ Nov. 1975) initiated by a leader circulating between clothes and skin. These pathognomonic figures, which follow the pattern of current flow, do not become white on pressure and disappear after one or two days. The lightning current also burns hair.

The current flowing through the body leads to electrification, an unpleasant but not deadly phenomenon. Electrocution (by ventricular fibrillation or asystolic arrest) is irreversible and

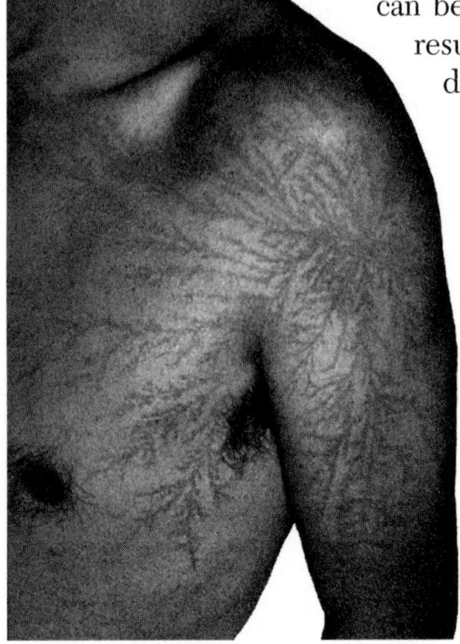

FIGURE 27. LICHTENBERG FIGURES ON HUMAN SKIN

can be deadly if cardio-pulmonary resuscitation (CPR) is not immediately administered.

Any lightning phenomenon is associated with a blast effect, leading to trauma (oedemas, contusions) by splattering or falling down (sometimes from elevated points). These barotraumatic effects can lead to internal hemorrhages and ruptured eardrums.

Numerous neurologic and sensory injuries have been documented: painful syndromes, paresias, paresthesias, hemiplegias, tetraplégias, paraplegias, irregularities of the cardiac rhythm, including fibrillation, myocardial infarction, speech irregularities including aphasia, loss of balance, amnesias, anxieties, cephalgias, sleeping difficulties and troubles, memory and concentration difficulties, depression, thunderstorm aversion, and various psychosomatic complaints.

Principal Reference

C.J. Andrews et al., *Lightning injuries: Electrical, Medical and Legal Aspects*, CRC Press, Inc., USA, 1992.

7
Secondary Effects and Lightning Damage

Side Flashes

In open air, living beings are not only sensitive to direct strikes (more likely when they are standing up), but also to side flashes, induced discharges, step voltages, and touch voltages.

It is dangerous to stay under an isolated tree (*see figure 28*) because if the body is fewer than several meters from the trunk, it may experience a side flash at the head or shoulder level.

Induced Discharges

Generally, all non-protected structures have to be avoided. It is better not to use small isolated structures (i.e., huts and barns) or buildings with metallic roofs isolated from ground, for example, supported by non-conducting poles as shelters from lightning, because a discharge can be induced by capacitive coupling (*see figure 29*).

Step Voltages

When lightning strikes the ground, the lightning current is spread out through the various layers of the soil. A high potential rise occurs at the point of strike. A step voltage can be experienced near this point. This can be especially troublesome for quadrupeds that can be electrocuted because of a potential gradient applied between anterior legs and posterior legs (*see figure 30*). In the latter case, electric currents cross the body passing through the heart. There have been many recent reports of lightning incidents involving sheep and cattle.

Figure 28. Don't stay under an isolated tree

Figure 29. Don't stay under metallic structures when these are not earthed

If caught in the open, the ideal position to be assumed is illustrated by the little girl in figure 30. Human beings standing up with joined feet can be hit by a direct strike. When walking (*see figure 30 on the right*), they can experience an electrification by step voltage; this situation is more hazardous if the ground resistivity is higher, the distance to the strike point is smaller, and the distance between the two feet is larger.

Touch Voltages

In a lightning storm, stay far from metallic structures, not only to avoid side flashes, but also to reduce risk of electrification by touch voltages. Indeed, electrization

FIGURE 30. ELECTRIZATION BY STEP VOLTAGE AND BEST POSITION TO BE ADOPTED IN OPEN AIR

by touch voltages (or contact voltages) occurs when people with feet in contact with the sufficiently conductive ground touch a conductive structure that may be at a different potential due to a lightning strike to that structure or its lightning protective system. Avoid touching metallic objects during thunderstorms (*see figure 31*).

FIGURE 31. DON'T TOUCH METALLIC OBJECTS DURING A THUNDERSTORM

Basic Safety Rules

Don't walk close to rivers and, more importantly, don't swim during a thunderstorm. Avoid horseback riding, bicycle or motorcycle riding, convertible cars, tractors, and harvesters (many farmers have been struck by lightning in the USA!). Nevertheless many lightning-strike victims do survive! According to the Guinness Book of Records, Roy C. Sullivan, an old forest ranger, has been struck seven times by lightning, from 1942 to 1977. He was given the nickname of Virginia Lightning Rod or Human Lightning Rod. He supposedly lost a toenail and his eyebrows, had his hair burnt, suffered various burns on his arms, legs, chest and stomach, but he survived! Unfortunately, disappointed in love, this famous lightning survivor committed suicide in September 2003. Perhaps the psychological effects of being struck by lightning also played a role in this tragedy.

In the countryside, walk far from the highest points, don't stay in a group, and avoid isolated trees. In urban environments move away from streetlights, towers, and metallic fences as well as isolated trees.

Suspend your golf game! Avoid open-air sports on fields, especially on the edge of a wooded area (where the potential gradient is larger) or near high metallic structures.

In a camping tent or trailer, be sure that a metallic conductor surrounding the area to be protected is properly grounded.

Keep away from water sports: no surfing, canoeing, or yachting, unless they are adequately protected by means of external metallic structures used as lightning rods extending into the water to provide electrical connection to ground potential.

In the mountains, run away quickly from the top and stay far from walls, cracks, crevices, edges, protuberances, and trees. Get rid of metallic objects. It is better to curl up in order to decrease the body surface and to protect yourself against diverted currents as shown in figure 30.

In town, rush into a store or a public building where the structure of the building is likely to provide an approximation to a Faraday cage.

If you cannot avoid being out in the open, exposed to a thunderstorm, take short steps or run (in which case only one foot touches the ground) and avoid holding any protruding metallic objects such as umbrellas, hiking poles, etc.

At home (*see figure 32*), it is best to take a position far from any chimney, and to unplug electrical devices and television antennas or telecommunication cables before the storm arrives (even when these cables are underground!).

Do not use a corded telephone. Keep clear from electrical lines, telecommunication lines, water or gas metallic pipelines,

FIGURE 32. BE CAREFUL INSIDE A HOUSE AS WELL!

as well as household electrical appliances (i.e., extractor hoods, dishwashers, electric heaters). Don't take a shower or bath during a thunderstorm.

A metallic car constitutes good protection as it serves as an excellent approximation to a Faraday cage.

The metallic skin of aircraft provides a good approximation to a Faraday cage. Scientists have shown that airplanes and rockets initiate lightning discharges from their extremities. On November 9, 2004 for example, two planes were struck by lightning when taking off and had to make emergency landings at Nice airport but they survived.

When somebody close to you is struck by lightning and is rendered unconscious, immediately perform CPR. This technique has saved many people.

Damage Due to Lightning!

In western Europe, lightning is responsible for more than fifty deaths per year and hundreds of injuries. Moreover, lightning damages or destroys thousands of houses, bell towers, chimneys, and tens of thousands of electrical and electronic appliances (telephone exchanges, computers, alarm systems, televisions, traffic signals, electric power plants, etc.).

One century ago, the relative number of injured people due to lightning was at least ten times higher than nowadays, since many people worked in open air, walked unprotected, and were ignorant of the elementary rules of prevention that we now understand. In France, newspapers did not hesitate to make the headlines and the front page to illustrate lightning strikes; see *Le Petit Parisien*, August 18, 1901 (*figure 33 on the left*) and *Le Petit Journal*, September 11, 1910 (*figure 33 on the right*) with a typical case of a bell-ringer inside a church bell tower.

In *Le Petit Journal*, on July 1, 1893, one can read an article with the heading: "Fontainebleau disaster":

The rain that people insisted to happen should have arrived more discretely and without hail and injuring lightning. Some days ago, at Héricy, close to Fontainebleau, three farmers, threatened by lightning, took an unwise step to take refuge under an enormous walnut tree. Lightning struck! One man was directly killed; his friend remained paralyzed on his whole left side. As for the woman, she was thrown down to the ground and only experienced some contusions. People were filled with consternation because of this misfortune. We wanted to describe

FIGURE 33. EXAMPLES OF FRENCH PEOPLE STRUCK BY LIGHTNING, ONE CENTURY AGO (FROM *LE PETIT PARISIEN* OR *LE PETIT JOURNAL*)

this case very carefully, hoping that our readers will stop seeking shelter under large trees when a thunderstorm is present, these large trees acting as hazardous lightning rods, intercepting the powerful lightning as Franklin devices do.

During the twentieth century, lightning appears less and less harmful. This is likely the result of urbanization and the common use of modern transportation means with metallic structures playing the role of a Faraday cage.

Nevertheless, many injured people and much damage have been reported. Here are some examples:

direct strikes to isolated people: France, summer 2002, one hiker (Hautes-Pyrénées), one mountaineer (massif de Néouville, Alpes), one jogger (Ile de Ré), and three walkers (two at Cap-d'Agde and one at Orange);

strikes to people in a group: from September to December 2003, the Democratic Republic of Congo was particularly

involved with thirteen people struck on the Kisangani market and eleven children killed in a Bikoro school (85 other injured children survived);
strikes to an isolated tree: a man thought he was protected under a tree at Tongeren in Belgium, but he died on July 18, 2004;
strikes in the mountains: one hazard among many other recent ones, two young men killed on September 9, 2005 on a peak dominating Schaffhouse in Switzerland;
strikes on water: lightning kills two canoe rowers on the Volga River close to Tsaritsyn in Russia;
quadrupeds killed with step voltages: the world record for domestic animals belongs to Denmark: on August 19, 2004 thirty-one cows were killed at the same time on Jutland Isle; elk herd of fifty-three animals was killed by lightning in Colorado in 1999;
forest fires: lightning ignites at least forty forest fires in Quebec and in British Colombia (Canada), in the summer of 2004;
blasts of bell towers: Rheinau Saint-Nicolas Church in Switzerland on August 6, 2004, Valflaunis Church in Herault (France) on October 14, 2004, Saint-John Church in Tournai (Belgium), on August 19, 2005;

FIGURE 34. EXAMPLES OF ROOFS AND FACADES DAMAGED BY LIGHTNING

roofs blown up by blasts, very often associated with fires: six houses caught fire in Var and Vaucluse (France) on September 4, 2002, a detached house Savigny-sur-Orge (Essone, France) on August 19, 2004, a primary school at Novalaise (Savoy, France) on September 15, 2004 (*see figure 34*).

Today, the most costly material damages are related to numerous electronic devices, but these damages are rarely recorded even by insurance companies.

These examples justify the importance of an effective protection against numerous effects of lightning. We elaborate on this crucial point in the following chapters.

References

NLSI (National Lightning Safety Institute), *Lightning Safety, with Risk Management of the Hazard, 2008, www.lightningsafety.com.*

V.A. Rakov and M.A. Uman, *Lightning – Physics and effects*, Cambridge Univ. Press, 2003.

PART IV
Lightning Protection

WHAT can one do to be effectively protected against lightning? Sometimes it may be difficult and costly, but adequate protection is possible by applying appropriate rules and standards, such as those issued by Technical Committee 81 (TC 81: Lightning Protection) of the International Electrotechnical Commission (IEC). In Europe, the Technical Committee TC 81X of CENELEC follows the same rules. These standards (IEC 62305) issued in 2006 (*see references*) are based on the concept of risk management in order to reduce damage caused by lightning significantly.

There exist neither devices nor methods capable of preventing lightning discharges. Direct or nearby lightning strikes to structures are hazardous to people and to the structures themselves, their contents, and installations as well as to services. That is why lightning protection measures have to be applied, taking into account a risk evaluation and management.

There is actually very little that is new regarding external lightning protection since Franklin's invention of lightning rods (1752) and the application of Faraday cages. The design of an external lightning protection system (ELPS), a better name than lightning rod which makes one think of a single vertical rod, has to be carefully considered from beginning of the design of a new structure in order to take into consideration the conductive elements of the structure including its grounding, where particular care must be taken in order to minimize the electric resistance.

Nevertheless, generally, an external lightning protection system alone is not sufficient to prevent damage to equipment from transients and surges caused by nearby lightning or direct strikes to the protection system of a building. Services (e.g., power lines, telecommunication lines) entering the building or structure and the protection against lightning electromagnetic pulses (LEMP) must be taken into account. Surge protective devices (SPDs) and electromagnetic shielding have constantly improved thanks to improved knowledge of the lightning current characteristics. SPDs are installed in parallel with an electrical circuit, acting essentially as an open circuit under normal conditions and becoming highly conductive under lightning over-voltage conditions.

This fourth section will provide practical information about lightning protection issues, so that the reader will at least be aware of simple rules and appropriate methods of protection.

8
Lightning Attachment and Structural Protection

The Attachment Process

The installation of simple vertical lightning rods (Franklin type) has been considered the key protection against lightning for a long time. The concept of a protection zone or cone was instituted as a guide to the deployment of lightning rods. The cone of protection is a conical volume centered at the rod tip with a fixed top half-angle, e.g., 30°, 45° or 60°, depending on the degree of protection desired.

It is important to note that some strikes terminate below the top of tall towers, antenna masts, and transmission line towers. A systematic observation of lightning strike points was conducted on very tall (> 500 m) television towers in Moscow and in Toronto.

In 1972, the Russian experimentalist Gorin wrote that the Moscow TV tower, 537m high (*see figure* 35), was struck 143 times in four-and-one-half years (average: thirty-two times per year) with a maximum of twelve times during the same thunderstorm. Eighty-three strikes were photographed. The maximum value of the peak current was 46 kA (kiloamperes). The front and tail times of the current wave shapes varied between 1 and 10 μs (microseconds) and between 20 and 70 μs, respectively. Of the forty-nine upward flashes recorded, most of the flashes were initiated from the top, but several struck the tower between 12 and 36 m from the top. Two downward flashes attached to the tower about 200 m and 300 m from the top. At

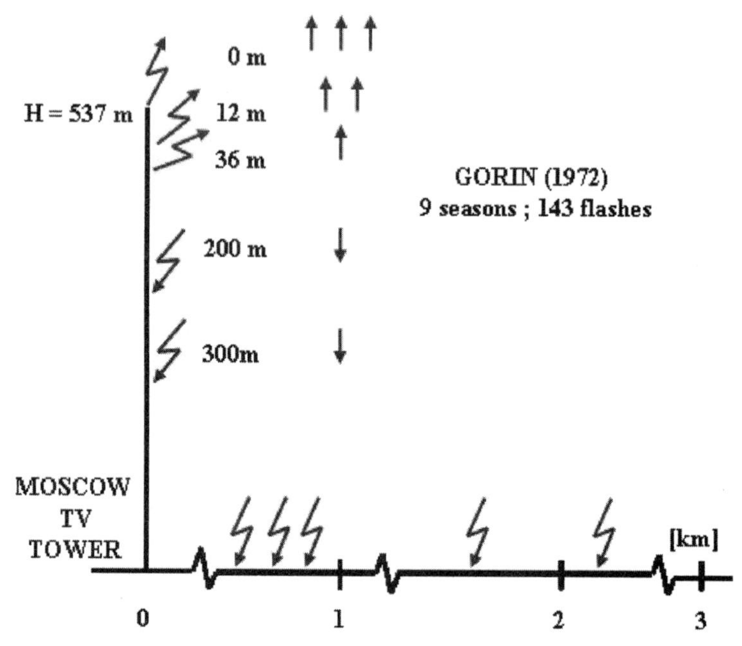

FIGURE 35. LIGHTNING STRIKES ON THE OSTANKINO TV TOWER IN MOSCOW, RUSSIA (1968–1972)

distances exceeding 200 m from the axis of the tower, ground flash density can be assumed to be equal to that for a flat terrain (in the absence of the tower).

DOES A LIGHTNING ROD DISCHARGE A THUNDERCLOUD?

Originally, Franklin thought that the lightning rod would silently discharge a thundercloud and thereby would prevent the initiation of lightning. Later, he stated that the lightning rod had a dual purpose: if it cannot prevent the occurrence of lightning, it offers a preferred attachment point for lightning and then a safe path for the lightning current to ground. It is in the latter manner that lightning rods actually work; a lightning rod cannot appreciably alter the charge in a cloud. However, with only a single short lightning rod, structural lightning protection is illusory. The key to effective protection is to form an approximation to a Faraday cage, unless the structure is metallic.

Radioactive lightning rods have no better efficiency than classical lightning rods with the same geometry. They are now forbidden in most countries.

DOES A LIGHTNING ROD ATTRACT LIGHTNING?

Not really. A descending lightning leader is influenced by a grounded object only when it comes within 100 m or so of that object. Thus, a lightning rod (external lightning protection system) intercepts imminent lightning strikes and provides a safe path for the lightning current to ground, but does not attract lightning from appreciable distances.

For about thirty years, various so-called active rods equipped with devices claiming to improve rod performance appeared on the market. Among these rods, the Early Streamer Emissions (ESE) type involves devices with piloted corona effect or devices generating sparks. Nevertheless, so far, there is no evidence that so-called active rods show any better efficiency than the classic Franklin rods of the same geometry.

IS AN ACTIVE ROD MORE EFFICIENT THAN A CONVENTIONAL ONE INSTALLED UNDER THE SAME CONDITIONS?

Not a single laboratory test or *in situ* test has ever demonstrated that a so-called active rod (i.e., repeller, ionizing, ESE, dissipation array, eliminator) provides better protection than a classic rod of the same geometry. Thus, there is no scientific evidence that so-called active protection methods are more effective than conventional lightning protection.

Knowledge of the lightning attachment process has allowed scientists to develop an engineering model describing interaction of lightning with a given structure. This model is called the electrogeometric model or rolling sphere model (sometimes called fictitious sphere model).

The Electrogeometric Model

A downward stepped leader initiated from the negative charge region of a *cumulonimbus* carrying a negative charge proceeds to ground and increases the electric field at the ground surface until it exceeds the critical value for the initiation of one or more upward connecting leaders. The distance between the descending leader tip and ground (or grounded object) at the instant of initiation is referred to as the striking distance. One of these upward leaders (charged positively) meets the downward leader and gives rise to the return stroke that traverses the ionized leader channel back up to the cloud.

The vertical component of the electric field near ground depends on the amount and distribution of charge on the leader channel and on the distance separating the leader tip from the ground. It is thought that the leader charge is related to the prospective return-stroke peak current.

From this hypothesis, a relation was established linking the striking distance d and the return-stroke current peak I.

The electrogeometric model is based on this dependence. It does not discriminate among various types of strike object (e.g., ground, tree, building, lightning rod) or among different heights. Strictly speaking, this model is only applicable to flashes with negative downward leaders (90% of all cloud-to-ground flashes). The electrogeometric model is presently the best engineering tool for the design of external lightning protection systems. Some versions of this model take into account the type of strike object and its height.

According to the electrogeometric model, the striking distance d (in meters) is related to the return-stroke current peak (in kA) by the following relation (*see figure 36*):

$$d = 10\ I^{0.65}.$$

The grounded object, which first appears within distance d of the leader tip constitutes the strike point.

Though the electrogeometric model was developed for the negative downward leaders, it is sometimes also used for positive leaders. Even less is known about the attachment process for positive CG lightning than for negative CG lightning.

As an example, let us consider figure 37, which illustrates the protection effect of a vertical rod of height h (h = 80 m) for two different values of the striking distance d_1 (15 m) and d_2

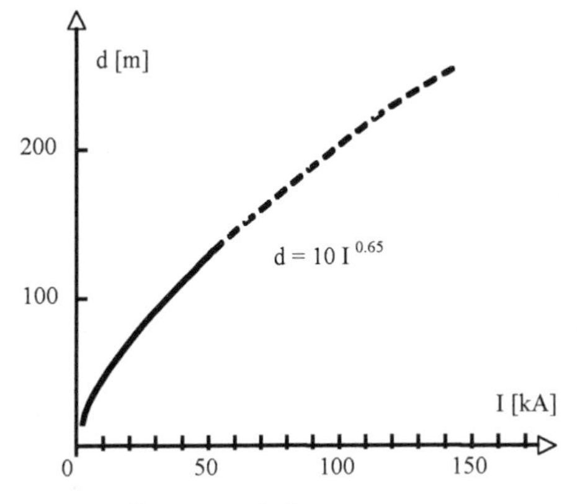

I [kA]	d [m]
2	16
5	28
10	45
30	91
50	127
100	(200)
150	(260)
200	(313)

FIGURE 36. STRIKING DISTANCE ACCORDING TO THE ELECTROGEOMETRIC MODEL

(100 m) corresponding to current amplitudes I equal to 1.9 kA and 35 kA, respectively. When the striking distance d is small (*see figure 37a where* $d_1 = 15$ *m*), the rod only protects a small curvilinear cone at its bottom and lightning can attach to the upper part of the rod and to the ground at radial distance exceeding $OA_1 = 15$ m; on the other hand, if the striking distance is larger than the rod height (*see figure 37b where* $d_2 = 100$ *m*), any object situated inside the big curvilinear cone will be protected and the lightning will attach to either the rod tip or the ground at a radial distance larger than OA_2. Note that the protection zone is larger when the current amplitude is larger.

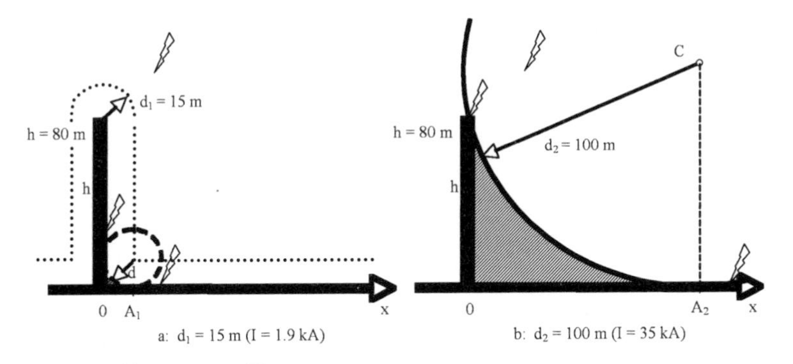

FIGURE 37. ELECTROGEOMETRIC MODEL APPLIED TO A VERTICAL ROD

DOES A SINGLE VERTICAL LIGHTNING ROD PROVIDE A CONICAL PROTECTION ZONE?

It was believed in the past that a lightning rod installed at the top of a high vertical mast would protect any object inside a conical volume centered on its tip with a top half-angle of 45° or even 60°. Previous national standards applied this rule. This is inconsistent (particularly for very tall structures!) with the observations and with the electrogeometric model, which is now widely accepted and included in most national standards.

Lightning Protection of Buildings and Other Structures

It is impossible to prevent lightning from occurring, as there is no device or method to avoid lightning strikes.

Cloud-to-ground discharges to or near structures can harm these structures, their contents and people. To minimize risk of injury and death as well as damage to buildings and their contents, lightning protection measures must be undertaken. How much protection is needed is determined using the risk management procedure.

Only a risk analysis allows the more or less objective evaluation of the required level of protection, which determines adequate protection means. The objective is to decrease the risk of damage or injury due to a direct lightning strike to the structure or inside the volume to be protected under a specific risk value called maximum tolerable risk. Damages depend on several characteristics, including the use, the content (persons and things) to be protected, the materials used and the protection measures employed to reduce the lightning risk.

The lightning current characteristics (*see chapter 4*) are chosen from data collected by CIGRE (*Conseil International des Grands Réseaux Electriques à Haute Tension*, the International Council for High-Voltage Electric Power Systems) and particularly wave shapes are relevant to components of short or large durations, possible components of downward flashes depending on polarity, components of upward flashes depending not only on polarity but also on the possible succession of different components. The wave shapes lead to specific design of lightning protection systems and electromagnetic pulses generated by lightning as well as tests on lightning protection devices.

Let N_d be the number of direct strikes to the structure per year (linked to the ground flash density N_g and to the collection area for strikes to the structure) and N_c be the maximum tolerable number of direct strikes to the structure per year.

We define a global protection efficiency E as:

$$E = 1 - (N_c/N_d).$$

When $N_d < N_c$, E becomes negative, indicating that it is not necessary to install a lightning protection system, though it is always advisable as a precautionary measure.

On the other hand, when $N_d > N_c$, a lightning protection system should be installed according to the requirements for the protection level chosen.

In the international standard IEC 62305, four protection levels are defined (I, II, III, and IV). Each level corresponds to a set of construction rules. The global protection efficiencies for each level are 98% (level I); 95% (level II); 90% (level III); and 80% (level IV). To obtain a protection level higher than 98% (level I^+), some additional lightning protection measures have to be applied.

More precisely, for a given protection level, a set of minimum and maximum values of the lightning current is specified. Maximum values of the lightning current peaks are set at 200 kA (99% of flashes have currents below this value) for level I, 150 kA (98% of flashes) for level II, and 100 kA (97% of flashes) for levels III and IV. Minimum values of the lightning current peaks are linked to the application of the rolling sphere method (radius R) in the design of lightning protection systems. They are fixed at 3 kA (99% of flashes have currents exceeding this value; R = 20 m) for level I, 5 kA (97% of flashes; R = 30 m) for level II, 10 kA (91% of flashes; R = 45 m) for level III and 16 kA (84% of flashes; R = 60 m) for level IV. The probability of current to be between the minimum and maximum values is easily estimated, and a set of protection measures is defined for this range of values by the protection level selected.

The efficiency of such protection measures is assumed to be equal to the probability for lightning current parameters to be inside this range.

IS A LIGHTNING ROD BETTER THAN OTHER AIR TERMINALS?

A single vertical rod cannot protect a big building! Multiple lightning rods are recommended by most lightning protection standards, but similar protective effectiveness is apparently obtained by covering the insulating roof of a building with a mesh of metallic conductors. The ultimate air terminal is a solid metal roof of adequate thickness to avoid lightning penetration at the strike point.

An external lightning protection system assures the efficient protection of persons, buildings, and other structures. The type and the positioning of the elements of the lightning protection system should satisfy the protection efficiency requirements, as well as the aesthetic ones, at as low as possible cost.

The external lightning protection system consists of:

- an air-termination system (horizontally meshed metallic conductors, vertical rods, overhead wires suspended on separate towers;
- a down-conductor system (generally vertical metallic conductors connecting the air-termination system to grounding electrodes or to measuring connections);
- for tall buildings, ring conductors of equipotential bonding (metallic conductors forming a horizontal loop around the structure assuring the electric equipotentialization of the down-conductor system);
- measuring connections (sections of down-conductors which can be dismantled to allow for the measuring of the grounding electrode resistance);
- an earth-termination system (buried metallic conductors assuring an electric contact with ground), vertical ground rods and a ring earth electrode (foundation earth electrode or loop connecting two or more earth electrodes, generally surrounding the structure to be protected).

Electrical continuity must be assured between different parts of lightning protective system by bolting or welding.

Some structural components can sometimes be used as parts of the air-termination system when they conform with the requirements for the elements of lightning protective system (i.e., electrical continuity, sufficient thickness).

The electrogeometric model is used for positioning air terminals according to the selected protection level (a rolling sphere radius equal to R = 20, 30, 45, or 60 m is used for protection levels I, II, III, or IV, respectively). Alternatively, one can install a square metallic mesh on the roof with the following mesh sizes: 5 m (level I); 10 m (level II); 15 m (level III); or 25 m (level IV).

The positioning of the exterior lightning protection system must be considered at the early stages of the design of a new structure, in order to take advantage of metallic parts of the structure itself, including its foundation.

For buildings taller than 20 m, down-conductors are interconnected with horizontal ring conductors separated vertically by 20 m.

Down-conductors are distributed around the structure so that their mean separation does not exceed 10 m, 15 m, 20 m, or 25 m, for protection levels I, II, III, or IV, respectively. A minimum of two down-conductors is required.

Three examples of lightning protection of typical structures are illustrated in figures 38 (villa); 39 (large building); and 40 (small industrial plant). Note that a ring-earthing electrode is employed in all three examples.

C. Bouquegneau and B. Jacquet showed (*see references*) that relatively short ground electrodes (vertical or inclined) interconnected by a ring-earthing conductor (foundation earth electrode) can be used to improve lightning protection (*see figures 38–40*).

SHOULD AN EARTH-TERMINATION SYSTEM INSTALLED FOR LIGHTNING PROTECTION BE SEPARATED FROM OTHER GROUNDING SYSTEMS?

No, all neighboring grounding systems in the immediate vicinity of each other (within several meters) should be interconnected

FIGURE 38. OPTIMAL LIGHTNING PROTECTION OF A VILLA

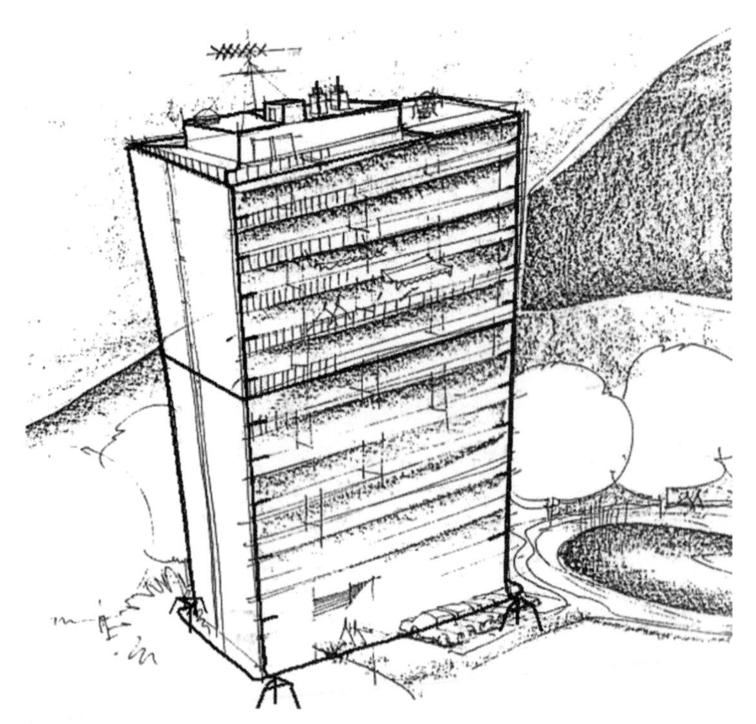

FIGURE 39. OPTIMAL LIGHTNING PROTECTION OF A LARGE BUILDING

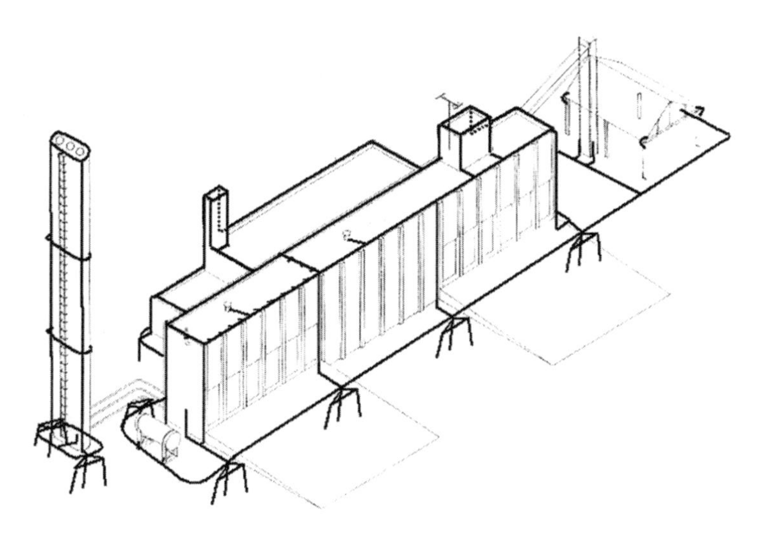

FIGURE 40. OPTIMAL LIGHTNING PROTECTION OF A SMALL INDUSTRIAL PLANT

to avoid potentially destructive electric arcs that may develop at or below ground level and in the structures to be protected.

References

C. Bouquegneau and B. Jacquet, *How to improve the lightning protection by reducing the ground impedances*, 17th ICLP, The Hague, The Netherlands, 1983.

International Electrotechnical Commission, IEC 62305 standard (2006), *Protection against lightning*, divided in four parts:
62305-1: General principles;
62305-2: Risk Management;
62305-3: Physical damages to structures and life hazard;
62305-4: Electrical and electronic systems within structures.

9
Beyond Structural Protection

Internal Protection

Generally an external or structural lightning protection system (*see preceding chapter*) is not sufficient to prevent damage to sensitive electronic equipment inside buildings. Care must be taken to protect services entering the structure, particularly electric lines and telecommunications lines, and to protect against LEMP (lightning electromagnetic pulses).

Here we are concerned with SPDs (surge protective devices) and magnetic shielding.

Electronic systems (e.g., data processing, telecommunications, control systems) can be subjected to transient over-voltages via different types of coupling. The IEC 62305 international standard mentioned above gives some information on the design, construction, control, maintenance, and laboratory testing of installations for the protection of electronic systems situated on or inside structures. This standard also gives suggestions for optimal protection efficiency in order to improve the collaboration between the system designer and the installer of the LEMP protection measures.

A lightning strike to a building can lead to destruction or significant disturbances in local electrical and electronic equipment due to:

- the high intensity of the transient magnetic field and the resultant induced voltages in the circuits;
- resistive coupling with the ground potential rise;
- electric-field coupling.

Wires and cables entering the struck building serve to transfer the local ground potential rise over some distance, possibly affecting other installations. These cables frequently bring induced surges. Their peak value is smaller but they are able to lead to significant damage, particularly when the equipment is very sensitive, such as data processing systems.

Some basic rules must be applied to the earth-termination system to minimize the potential for damage from transients and surges. All earth electrodes of the neighboring structures linked by electrical and telecommunications cables have to be interconnected. A meshed earthing system is better because such a system reduces the current amplitudes in the cables because there are numerous parallel paths available to the lightning current. The lightning earth-termination system and the low-voltage earth-termination system are to be bonded outside the building. Each down-conductor is to be connected to the foundation earth electrode, otherwise a ring-grounding electrode is necessary. Cables must be placed in metallic pipes or in steel-reinforced concrete conduits, which must be integrated with the meshed earth-termination system.

In order to reduce electromagnetic interference, magnetic shielding must be employed and all the large metallic parts must be connected to each other and connected to the lightning protection system. Efficient shielding of the equipment significantly reduces the magnetic field inside the equipment, in the 10 kHz (kilohertz) to 10 MHz (megahertz) range. When shielded cables are used inside the structure to be protected, shields must be bonded at both extremities as well as at each lightning protection zone boundary. Cables connecting two different structures must be placed inside metallic sheaths such as tubes, grids, grid-like reinforced concrete, bonded to equipotential bars of the structures. Cable sheaths have to be bonded to these bars. Electric conduits are useful when cable shields cannot convey the expected lightning currents without damage.

The IEC 62305 standard defines different lightning protection zones (LPZ) characterized by important changes in the electromagnetic conditions at their boundaries. However, these concepts are too technical to be discussed here.

Electronic equipment is particularly sensitive to surges. Transient low-energy signals entering equipment may destroy components and lead to a premature aging of its insulation. These surges can enter the equipment by different pathways such as signal inputs- and outputs, power connections, radiated electromagnetic fields, and grounding.

Systems that can be subjected to large electromagnetic impulses must be protected at all possible levels of penetration in order to prevent surges from reaching sensitive systems. These protection means are based on the following principles:

- to avoid any strike to systems and any current circulating in the equipment, installations, and bonding conductors between them;
- to limit the level of over-voltages induced in the conducting network of the building;
- to limit the electrical potential rise of the ground electrode as well as voltages between neighboring earthings (to connect all earthing electrodes to assure the same potential!);
- to prevent the penetration of over-voltages capable of destroying or causing malfunction of the system.

When Do You Have to Protect Yourself Against Lightning?

In 1769, Saint-Nazaire Church at Brescia (Lombardy, Italy), used as a warehouse with one hundred tons of gunpowder, was struck by lightning. The gunpowder exploded, killing more than 3,000 people and destroying a large part of the city.

In modern times all warehouses with explosive or inflammable materials are supposed to be protected against lightning at a level I^+ (i.e., at a level I with additional measures). In this case, a double Faraday cage is imposed (the second cage being built at a minimum distance of 2 m from the first one).

Lightning risks leading to the possibility of chemical pollution are considered to be serious. A lightning protection system is needed when there are risks of injury to human beings, loss of service to the public, loss of cultural heritage or loss of economic value due to non-delivery of promised services and downtime on the job (these four types of losses are considered in the IEC 62305 standard).

In the risk assessment, several parameters must be known: the keraunic level or the lightning flash density, the lightning collection area of the structure to be protected (equivalent capturing area, i.e., the actual ground area plus a surrounding band of width equal to three times the height of the structure, if it is smaller than 30 m in height), and height and contents (e.g., hazardous materials) of the structure to be protected.

The IEC 62305 standard considers three types of possible damage: D_1, injury to living beings due to step and touch voltages (*see chapter 3*); D_2, physical damages (e.g., fires, explosions, chemical pollution, mechanical destruction) due to electrical discharges and effects of the lightning currents; D_3, failures of electrical and electronics systems coming from overvoltages. Four types of sources of damage are also introduced: S_1, flashes to a structure; S_2, flashes near a structure; S_3, flashes to a service entering the structure; and S_4, flashes near a service entering the structure.

Four types of risks correspond to equivalent losses: L_1, loss of human life in a structure; L_2, loss of service to the public; L_3, loss of cultural heritage; and L_4, loss of economic value (e.g., structure and its contents, services, suspension of activity), the last component (economic value) being left to evaluation by a national committee.

Today, physical damages leading to the most expensive repairs occur in electronic systems; unfortunately, even insurance companies rarely index these damages.

Basic protection criteria comprise:

- protection against fires and explosions as well as losses of life by means of an external lightning protection (air-termination system, down-conductor system, and earth-termination system) and an internal lightning protection (equipotentialization and minimum separation distances between metallic objects subjected to different electrical potentials);

SOME PROBABILITIES OF LIGHTNING STRIKES

First case: isolated rectangular villa (15 m × 10 m), 5 m high, in a region where the lightning flash density is equal to 1.2 $km^{-2}.year^{-1}$; the collection area for this relatively low building is 1.6×10^{-3} km^2, the probability of lightning strike would be 0.00192 flashes per year, i.e. one flash every 520 years! Generally, in this case, no protection is needed.

Second case: an adult 1.83 m high standing permanently on the same site as in the first case, for example, is expected to be struck by lightning once every 7,350 years!

Third case: a rectangular chemical plant (200 m × 50 m), 30 m high, in a region where the lightning flash density is equal to 3 $km^{-2}.year^{-1}$, with a collection area equal to 0.08 km^2, is expected to be struck by lightning once every four years. If plant materials are considered hazardous, such a plant has to be protected against lightning and the risk assessment would confirm that the highest level of protection (level I^+) is needed.

- protection against lightning electromagnetic pulses (LEMP) (definition of lightning protection zones with shields, equipotential bonding, and grounding);
- protection of services entering the structure to be protected, namely telecommunication lines (selection of line components with adequate characteristics, selection of convenient magnetic shielding, use of surge arresters).

All lightning protection installations must be maintained and inspected by experts or qualified technicians.

Without any risk calculation, we can say that a lightning protection is not necessary in some cases. In many cases, modern houses in urban regions, if these houses are surrounded by many neighboring ones of equal or exceeding height may not need explicit lightning protection. However, only a complete calculation of the risk, taking into account all the parameters indicated above can justify such a decision (a good way to proceed is to use the RISK Multilingual software, based on the application of the IEC 62305-2 standard, which can be ordered by e-mail at Christian.bouquegneau@fpms.ac.be).

Lightning Protection of High-Voltage Transmission Lines

Since the early days of three-phase high-voltage transmission lines, because of tall metallic towers, failures due to lightning over-voltages have been a principal cause of service interruptions.

Towers and conductors of transmission lines are preferred targets for lightning. Shield wires (connected to grounded towers) were added above the high-voltage transmission line conductors for the latter to be protected (*see figure 41*). These wires are also referred to as overhead ground wires.

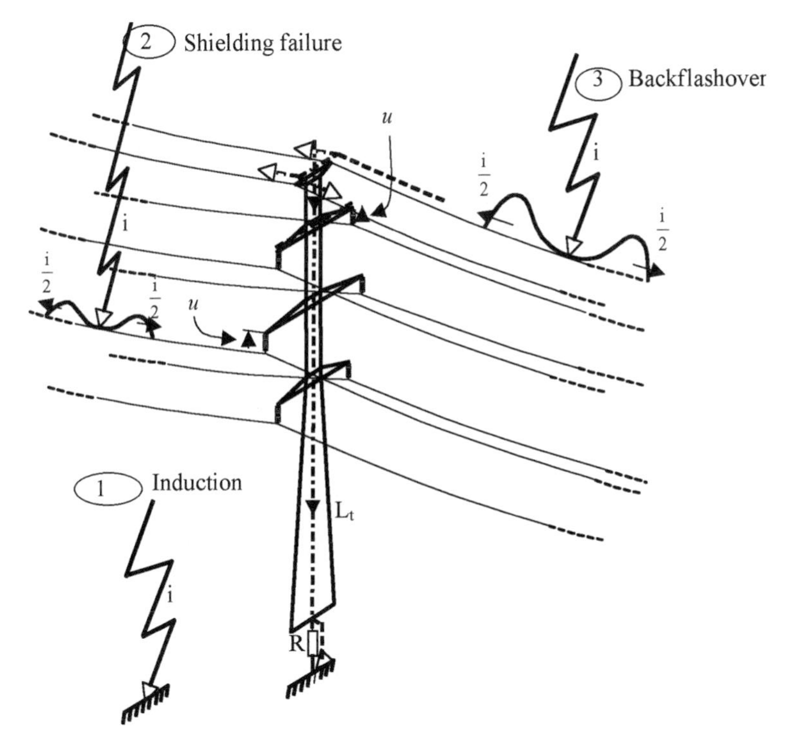

FIGURE 41. THE THREE MODES OF LIGHTNING INTERACTION WITH HIGH-VOLTAGE TRANSMISSION LINES

Three mechanisms can be responsible for the flashover of insulator strings:

- induction flashover (1), when lightning strikes ground near the transmission line without direct current injection into any of its elements; however, this mechanism is less harmful for higher voltage power lines. This mechanism is important for lower voltage (<35 kV or so) power distribution lines.
- shielding failure (2), when lightning strikes a phase conductor, which means that the ground wires failed to shield (intercept the strike);
- even when there are shield wires whose function is to eliminate direct strikes, the transmission line is not always completely protected; indeed, when lightning strikes a shield wire (or the tower itself), the lightning current flows to ground through the nearest tower, increasing the tower potential, especially when the ground resistance is very high and/or the tower is very tall and lightning current is

a b

FIGURE 42. "LIGHTNING-PROOF" *BOUQUEGNEAU* TOWERS FOR HIGH-VOLTAGE TRANSMISSION LINES:

A. ANGLE TOWER (EAR-SHAPED CORNERS)

B. STRAIGHT LINE TOWER

very steep. If the amplitude of voltage between the tower cross arm and phase conductor exceeds the critical flashover voltage of the insulator string, a secondary flashover or backflashover (3) can produce a failure similar to the one produced by direct strikes to phase conductors.

By applying the electrogeometric model, specifically the version designed by Whitehead for high-voltage transmission lines, one of the authors (C.B.) was the first in the 1970s to propose a "lightning-proof" tower with negative protective angles. Shield wires are laterally exterior to the phase conductors (*see figures 42a and 42b and the reference at the end of this chapter*) for the classical flag-type towers of the Belgian network of high-voltage (150 kV and 380 kV) transmission lines. This tower is lightning-proof not only with respect to the shielding failure but also to the backflashover problems. Negative protective angles are usually recommended for very tall double-circuit transmission line towers.

The 380 kV transmission line with "lightning-proof" towers from Massenhoven (Antwerp, Belgium) to Maasbracht (The Netherlands) was built in the 1980s. Very low rate of failure has been recorded on this transmission line.

References

C. Bouquegneau and C. Grégoire, *The lightning performance of a suboptimal double-circuit tower for high-voltage transmission lines*, Electric Power Systems Research, New York, vol. 4, 1981, pages 159–165.

International Electrotechnical Commission, IEC 62305 standard (2006), *Protection against lightning*, divided in four parts:
62305-1: General principles;
62305-2: Risk management;
62305-3: Physical damage to structures and life hazard;
62305-4: Electrical and electronic systems within structures.

PART V
Concluding Remarks

IN the second part (chapters 3 to 5), we presented the state of the art in lightning phenomenology.

Briefly, lightning is a transient electrical discharge lasting, on average, some tenths of a second and whose length is in kilometers.

The origin of lightning is typically an electrically charged cloud (an electric dipole with an upper positive charge and a lower negative charge or a tripole). A flash from cloud to ground is only one of four possible types discharges (cloud to ground, cloud to air, intracloud, and intercloud).

In this last chapter, we discuss relatively new research on luminous events that occur in the middle atmosphere above the cloud tops and below the ionosphere and other topics that should be studied to improve our knowledge of the amazing natural phenomenon that is lightning!

We conclude with an answer to the question selected as the title of this book: **how dangerous is lightning?**

10
New Frontiers

Elves, Sprites, Blue Jets

Above thunderclouds, between tropopause and mesopause (*see color insert*), transient luminous events, most of which are associated with highest-intensity lightning discharges, still await for a complete scientific explanation. These are reddish elves, red sprites, and blue jets.

Red sprites were discovered in the United States in 1989, as a result of tests of a new low-light video system. They have durations of 10 to 300 milliseconds and have diameters from 1 to 50 km. Some researchers hypothesized that they could be associated with avalanches of runaway electrons (electrons moving at speeds near the speed of light, getting more energy from the ambient electric field between collisions with air molecules than the energy they loose in each collision) with energy higher than 1 MeV (one million electron-volts) or so. These avalanches could be triggered by cosmic rays in the mesosphere and the stratosphere. They interact with ambient air molecules and produce X-rays and secondary gamma rays.

There are also conical blue jets, up to 50 km in altitude, and disk-shaped elves, up to 100 km in radius. Elves expand radially at high speeds and last some milliseconds.

For some years in France, CEA (Atomic Energy Commission) and CNES (National Center of Spatial Studies) made a lot of observations thanks to the International Space Station in order to determine the energy emitted by these events and to record statistical data on the frequency of occurrence and their global distribution. A microsatellite, called Taranis (Tool for the Analysis of RAdiations from lightNIngs and Sprites), will be launched in the near future to observe these phenomena as closely as possible.

Similar experiments are pursued in the United States and by ESA (European Space Agency), in cooperation with the Danish National Space Center, observing the giant discharges in the middle atmosphere by means of cameras installed on top of mountains. An ASIM (Atmosphere-Space Interactions Monitor) monitor should soon be installed on the Space Station. The purpose is to identify coupling mechanisms between neutrosphere, ionosphere, and magnetosphere, above the tops of the thunderclouds that give rise to these spectacular events.

Do the transient luminous events alter the chemical composition of our atmosphere and do they play a role in determining our climate? Do they influence the ozone layer? We should find an answer to these important questions in the near future.

Laboratory Discharges Compared with Natural Lightning

We already mentioned that there are four types of ground discharges, depending on the polarity (negative or positive) and the direction of the initial leader (downward or upward). There are also bipolar flashes. Upward flashes are generally launched from tall structures (towers, chimneys, antennas, tall buildings) or from lower objects on top of mountains or hills.

In chapter 5 we saw that it is possible to initiate lightning artificially by launching small rockets that unwind a metallic wire attached to ground (at ground potential) toward the overhead thundercloud. Even though there is no first stroke in triggered lightning, these experiments have made it possible to determine many properties of lightning that could not be studied in natural lightning due to its random occurrence in space and time.

Simulation of the lightning channel in a high-voltage laboratory has very limited application, since it does not allow the reproduction of many lightning features and it does not allow the testing of large distributed systems.

FIGURE 43. HIGH-VOLTAGE LABORATORY OF THE POLYTECHNICAL UNIVERSITY OF MONS (BELGIUM) (ON THE LEFT, A 1 MV CASCADE TRANSFORMER AND ITS CAPACITIVE VOLTAGE DIVIDER; IN THE MIDDLE, 2.4 MV MARX IMPULSE GENERATOR AND ITS RESISTIVE VOLTAGE DIVIDER; ON THE RIGHT, SPHERE-TO-SPHERE GAP FOR CALIBRATING THE APPLIED VOLTAGE)

A number of new approaches to triggering lightning, including those using lasers, microwave beams, water jets, and transient flames, have been proposed, but it appears unlikely that any of these "non-conventional" approaches will lead to an operational technique in the near future.

Research in Progress

Our knowledge of the physics and effects of lightning on living beings and structures still needs improvement. Numerous questions are waiting for an answer to be provided by the world scientific community.

For example, mechanisms of the formation and electrification of thunderclouds are not yet completely understood. We do not have enough experimental data on natural lightning, particularly on its in-cloud development and its attachment process. We do not understand yet the physics of initiation and propagation of the first component of a negative ground discharge and even less the phenomenology and mechanisms of positive lightning.

Among the most principal questions awaiting urgent scientific answers are the following:

- How is lightning initiated inside a *cumulonimbus* cloud?
- What are the different physical processes of a lightning flash?
- How can properties of a lightning flash be deduced from electromagnetic radiation measurements?
- What are the development mechanisms of stepped leader and dart leaders?
- What is the attachment process for various objects?
- How does a return stroke develop?
- What are the production mechanisms of X-rays and gamma rays emitted by lightning discharges?
- What is the mechanism responsible for the production of various gases in the atmosphere by various lightning processes?
- How can we observe ball lightning and how can we explain its mechanism?
- What is the relation between lightning and climatic phenomena?

There is still a lot of work for the fulminologists all around the world, due to the high variability of lightning characteristics from one location to another.

Many lightning properties and mechanisms remain a subject of debate. As an example, for the lightning initiation and preliminary breakdown, some scientists favor the theory of positive streamers that are developing from hydrometeors, others prefer a mechanism that invokes bipolar streamers coming from a chain of precipitation particles, still others insist on the role of runaway electrons.

Stepping is observed in negative leaders regardless of whether they are initiated in the cloud (downward leader) or at the grounded objects (upward leader). Similar stepping also occurs in laboratory long sparks (with a maximum length of the order of 10 m, while the typical step length in lightning is about 50 m).

The physics of lightning attachment is still not clear. The attachment process begins when a downward leader induces upward leaders from ground or various elevated grounded structures. Generally, only one or two of these upward leaders connects with the stepped leader, causing the first return stroke. The stepped leader develops in a virgin (non-ionized) air, while subsequent strokes usually propagate inside a warm previously ionized channel.

Lightning strikes to aircraft are generally initiated by the nose or tail or the sharp extremities of the wings. There are attempts to model the sweeping of the lightning attachment point along the aircraft in flight.

Recent observations show that lightning produces extensive branched arcing along and below the ground surface.

Fulgurites are glassy tubes with zigzagged and branched appearance that are produced by the lightning current in sandy soil. The mechanism of lightning current dissipation in soil is not yet well understood.

Of all the lightning processes, the return stroke is the most energetic and most investigated. Physical properties of the return-stroke channel such as temperature, electron density, pressure, have been estimated from time resolved optical spectroscopy. However, there is no comprehensive model of the lightning return stroke that could predict most of the observed characteristics of the return stroke.

Runaway electrons created by the electrical discharge could produce X-rays (in the range from 30 to 250 keV for dart and stepped leaders) and gamma rays (in the MeV range) in their interaction with air particles. For the time being, the mechanism of runaway breakdown in lightning leaders (and in laboratory sparks) is not yet clear.

Atmospheric electric discharges, including corona and different processes in lightning discharge and transient luminous events (sprites, blue jets, and elves) between cloud tops and ionosphere, produce new trace molecules from the constituents of the atmosphere: nitric oxide (NO) is the most common, and it influences the production of ozone (O_3). Tropospheric ozone absorbs infrared radiation and is therefore acting as a greenhouse gas. Ozone in the stratosphere is important to life on Earth because it shields the Earth from the Sun's harmful ultraviolet radiation. There is a wide disagreement on the percentage of the contribution of NO_x in the atmosphere due to lightning, varying from 3 to 20% of the total.

Lightning is a highly variable event and no two lightning discharges are the same. Even though measuring the time varying electric and magnetic fields from lightning is relatively easy, estimating the spatial and temporal distribution of sources that give rise to these fields is not so easy. This type of problem is sometimes called the inverse source problem in lightning. As electromagnetic fields travel along the Earth they suffer from attenuation and dispersion due to its finite conductivity. Simple models that can roughly predict the peak currents in the lightning channel from remote measured fields are available. Simultaneous measurements of fields at different stations can be used to locate the point of lightning strike. Interferometric techniques can give information on the development and progression of lightning discharges inside the clouds. More accurate models are required to infer the properties of lightning processes from remote field measurements.

A strong need exists today to understand the way in which the lightning flashes interact with power systems, telecommunication systems, railway signalling systems, wind turbines, aircraft, and other special structures. This knowledge will lead to better protection measures to safeguard these man-made systems against lightning.

Lightning and Climate Change

Many researchers study relationships between lightning, rainfall, and severe weather. Data from the NASA Tropical Rainfall Measuring Mission (TRMM) satellite, launched in 1997, are used to investigate the relationship between rainfall and lightning. Among other instruments, this satellite has precipitation radar and a Lightning Imaging Sensor. Convective rainfall is well correlated to lightning activity, although the relationship varies from region to region. Improving the understanding of these relationships should allow the use of lightning data to estimate rainfall amounts for forecasting purposes. Downbursts and microbursts, the strong winds blasting down from thunderstorms are major hazards to civil aviation. The time of occurrence of microbursts is apparently closely related to the occurrence of lightning flashes in the thundercloud, at least in some cases, and a knowledge of the connection between them could aid in the prediction of the occurrence of microbursts.

It has been shown that increases in the positive cloud-to-ground flash rate often occur shortly before the onset of severe weather (tornados, strong wind, hail). Knowledge of the physics of lightning, especially the mechanism of initiation of lightning in thunderclouds, will help to link statistics of lightning occurrence and lightning characteristics with particular weather phenomena.

The global lightning (global electric circuit) studies are also related to the Earth's climate. Recent observations have shown strong connections between regional/global lightning activity and important climate parameters, such as surface temperature and upper troposphere water vapor. Global lightning activity could be used as a global thermometer to track changes in the Earth's climate. It has been suggested that a 1°C increase in global average temperature could produce 10% more lightning flashes.

The quantification of the contribution of lightning flashes in the production of nitrogen oxides in the atmosphere is an important piece of information in finding the effects of man-made nitrogen oxides on the global environment changes. This is related to recent debates on effects of pollution on the conservation of our beautiful planet.

Lightning research requires more and more interdisciplinary research teams, and the topics being studied are fascinating.

How Dangerous Is Lightning?

In a newspaper article published on December 2, 2005, we can read:

As the proverb goes, lightning never strikes twice at the same point. Nice saying for a Canadian family whose house was struck by lightning three times in six years. . . . According to the

Environment Canada Agency, the probability to be struck by lightning in Canada is one out of one million. . . . Far from the unusual appearance of the event, each time lightning strikes, it causes a lot of big damages in the house.

These are prejudices and wrong ideas (*idée reçue*)! If a house was damaged by lightning a first time, it needs to be protected against lightning according to the general principles described in chapter 4.

In a well-protected building, there is nothing to fear, if all the precautions discussed in chapter 7 are taken. The basic safety rules are applicable everywhere especially outside the buildings.

Man is no longer helpless facing this natural phenomenon. But we should remain very careful even if during the last century fewer people have been killed by lightning, due to urbanization, a better understanding of risks posed by lightning and a better knowledge of lightning phenomenology (*see part II*) and its effects (*see part III*).

Even if lightning continues to captivate and frighten, it seems to be less and less mysterious. Of course lightning gave birth to numerous myths from the beginning of all civilizations (*see chapter 1*). Unfortunately, these myths continue to live among numerous non-scientists.

In conclusion, we quote Heraclites, who said lightning (the Heaven Fire) governs everything. Let us remain a little more modest while bringing to the attention of people several facts: lightning occurs on our planet about one hundred times per second and is probably responsible for the origin of life on Earth. It is a natural fertilizer and it is a generator, which constantly restores the Earth's electric charge (*see chapter 4*).

It has been said that lightning is one of the essential forces acting on a world in continuous evolution.

Principal Reference

COST Action P18, *The Physics of the Lightning Flash and Its Effects*, technical annex, European Union document, European Science Foundation, June 2005.

Appendix
Electrical Discharges in the Laboratory

Electrical Breakdown

Electrical breakdown in gases, including air, is the change from the insulating state of the gas to its conductive state.

In very small gaps (distance d between electrodes of some centimeters), the breakdown mechanism is described by the well-known Townsend theory. In longer gaps (d equal to some decimeters), the theory of streamers must be considered. In uniform electric fields, the dielectric strength in air at STP is about 3 MV/m (atmospheric pressure).

Physically, applying an electric field to a gaseous dielectric first leads to a partial ionization (formation of positive ions and negative electrons). The exponential increasing of ionization generates *electron avalanches*. At a critical value of the electric field, these avalanches either lead to *streamers* that bridge the breakdown gap when this gap is sufficiently short or develop into *leaders* in longer gaps.

The electrical breakdown is thus facilitated by the development of streamers.

In long gaps, several streamers can develop from one point of the cathode. Heat generated by streamer currents increases the temperature at the head of the streamer. When the temperature is high enough, thermal ionization at the avalanche head generates electron detachment and a steep increase of the electron concentration that leads to increased electrical conductivity of the plasma: the streamer then develops into a leader. As the leader is a relatively good electrical conductor, the cathode potential is transferred to the leader tip, so that it can produce other streamers. The leader rapidly expands to the anode, and a spark is formed. When the current is large enough, the spark converts into an electric arc (or in other types of plasma).

Streamer Theory

In 1940, Craggs, Meek, and Raether elaborated the streamer theory, now completed by numerous essentially experimental contributions.

An electron avalanche initiated by a single electron exponentially multiplies the number of electrons at the avalanche head. These electrons diffuse and the avalanche extends in length, leading to an increasing of the electric field induced by the space charge. At a critical length, the induced electric field approaches the electric field applied externally. Then the electron avalanche is converted into a streamer.

According to Meek, three conditions are necessary for the occurrence of a streamer:

1) production of very energetic photons at the principal avalanche head;
2) ionization of gas molecules near the avalanche head;
3) sufficient space charge at the principal avalanche head so that adequate secondary avalanches can be generated in the reinforced electric field.

The mechanisms of formation of a positive streamer and a negative streamer are different (see V. Cooray, *The Lightning Flash*, IEE, Power & Energy Series 34, UK, 2003).

Let us consider the development of a *negative streamer* (anode directed streamer), i.e., a streamer initiated at the cathode and flowing to the anode (see figure A.1) in a uniform electric field.

The electrons of the avalanche move into the gap leaving behind positive charges close to the cathode. When the avalanche reaches the critical size, secondary avalanches extend the positive space charge towards the cathode. When the positive channel reaches the cathode, both the field enhancement associated with

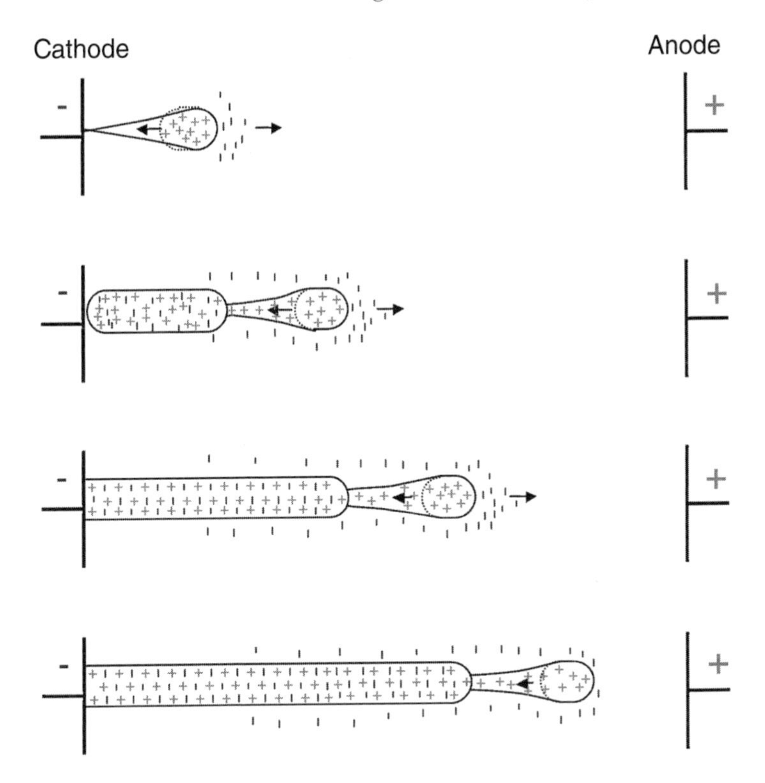

FIGURE A.1. DEVELOPMENT OF A NEGATIVE STREAMER

the proximity of positive space charge to the cathode and the collision of positive ions on the cathode lead to the emission of electrons from the latter. These electrons will neutralize the positive space charge generating a weakly conducting channel that connects the negative head of the electron avalanche to the cathode. The high electric field at the avalanche head pushes the negative space charge further into the gap while the positive space charge left behind is neutralized by the electrons supplied by the cathode and travelling along the weakly conducting channel connecting the streamer head and the cathode.

If the background electric field is very high, the positive space charge of the avalanche may reach the critical size necessary for streamer formation before reaching the anode. This may lead to the formation of a bidirectional discharge the two ends of which travel towards the anode and the cathode, the former as a negative streamer and the latter as a positive streamer. Such a discharge is called a *mid-gap streamer*.

The avalanche-to-streamer transition takes place when the number of charged particles at the avalanche head exceeds a critical value N_c. Raether and Meek estimated that an avalanche will convert to a streamer when the number of positive ions in the avalanche head reaches a critical value of 10^8.

A streamer crossing the gap does not necessarily mean that it will always lead to electrical breakdown of the gap. To do so, the channel must be heated sufficiently to become conductive enough. This fast heating of the gas (thermal ionization) is due to a thermalization process to obtain the streamer to spark transition.

In a plane uniform gap of separation d, a condition for the streamer to be initiated is that an avalanche length reaches the critical value equal to d. This corresponds to an electric field higher than 26 kV/cm, a higher value than the one needed for its propagation. Once streamer is initiated, it propagates all the way across the gap.

Electrical Discharge in a Non-Uniform Field

In a non-uniform field E (see figure A.2), the generated streamer must continue to propagate because of the applied electric field.

If the background electric field beyond x_c decreases below $E_c = 5 \times 10^5$ V/m, no negative nor positive streamer will reach the grounded electrode and the streamer will only progress on short distances from 0 to x_c, i.e., in the preferred ionization region. If the background electric field lies between 5 and 20 kV/cm, a positive streamer will cross the gap but a negative streamer may die out before reaching the anode. This explains why it is easier to cause breakdown in a rod-plane gap when the rod is at positive polarity than when it is at a negative polarity.

In a non-uniform electric field, space charges and electrode polarities influence the breakdown voltage. The latter depends not only on the product of the pressure p and the gap distance d but also of the shape factor taking into account the electrode geometry through an enhancement factor of the electric field f_r defined as the ratio of the maximum value E_{max} of the electric field (smaller

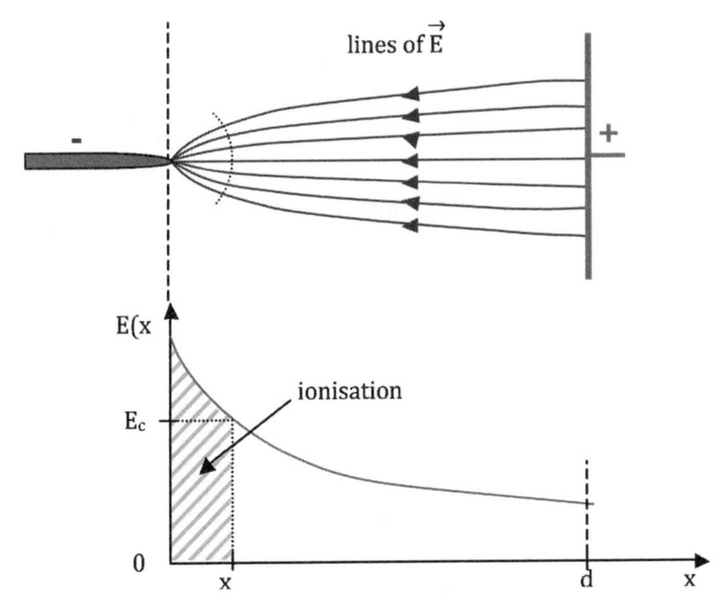

FIGURE A.2. STREAMER BREAKDOWN CRITERION IN A NON-UNIFORM GAP

electrode radius) to the average value E_{ave} (= U/d) in the gap. When f_r is smaller than 5, the electric field is close to uniform in the gap and the discharge occurs as if the field were uniform.

Space Charge Effects

Gaps with non-uniform fields are characterized by a discharge, which is always initiated from the electrode with a higher electric field (sharper geometry). Its polarity then plays a very important role.

Initiated discharges are not necessarily *complete*. They can be partial or incomplete, even at high voltages. However, an incomplete discharge can extend through the whole gap and induce a complete discharge. The incomplete to complete discharge transition generally occurs when d is of the order of 2.5 times the electrode radius from which the partial discharge is initiated. When partial discharges are initiated at the interface electrode-gas, they are called *corona discharges*.

Corona effect is so called because of the luminous glow appearing around, for example, cylindrical conductors of small radius when the electric field reaches a critical value.

In a *negative corona* discharge (see figure A.3), electron avalanches develop towards the anode in a decreasing electric field, a positive space charge being left behind close to the cathode. This space charge serves to slow down the electrons up to a critical distance r_0 where the electric field is smaller than its threshold value for ionization. At this distance, electrons are captured by oxygen molecules of air (electronegative gas); the negative ions formed generate a negative space charge. Both space charges of opposite polarities change the field configuration in the gap: the electric field increases further if we are closer to the cathode and decreases in the opposite case. Thus successive avalanches develop in higher electric fields but travel over shorter distances.

There are three types of corona discharges: *Trichel pulses*, non-impulsive *negative glows*, *negative streamers*.

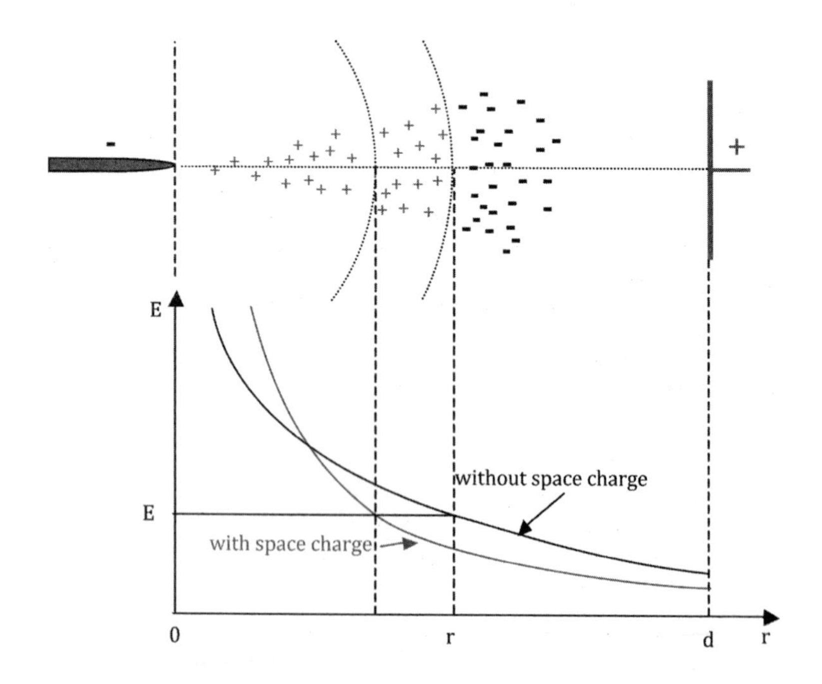

FIGURE A.3. CORONA EFFECT WITH A NEGATIVE ROD

1) The partial discharge disappears when the effective electric field decreases below the critical field. After the disappearing of the space charge, the resulting background field is higher and the cycle starts again. This process leads to pulses of corona current discovered by Trichel. They are called *Trichel pulses*. They last some tens of nanoseconds. Their frequency that depends on the cathode geometry and the gas pressure increases with the background field. The time interval between two successive pulses is from 1 to 100 μs. Trichel pulses do not occur in the absence of electronegative gases.
2) If the electric field is high enough to rapidly transfer the negative charge toward the anode, Trichel pulses do not appear. The discharge looks like a *negative glow*. Moreover, positive ions acquire enough energy from the electric field to bombard the cathode and to throw out a large number of electrons into the gas.
3) When the electric field is still higher, the generation of the space charge is such that the avalanches become real *negative streamers*. These partial discharges by streamers propagate far from the region where the electric field is weak into the gap. Their extension increases when the background field increases. The streamers generate low-frequency pulses in the discharge current.

Long Sparks

Long breakdown gaps are natural spaces where the discharges spread out over large distances (hectometers or even kilometers) or laboratory gaps that are more than 5 m long.

From time 0 of voltage applied to a gap and the following breakdown at time t, a total time lag t is the sum of a *statistical time lag* and a *formation time lag*. The statistical time lag is the time needed to generate a primary electron initiating the first avalanche. This time lag decreases when the cathode irradiation by ultraviolet rays is more intensive or when the applied voltage is higher. The formation time lag is the time necessary to develop the discharge from the first avalanche.

To come to breakdown after applying a pulse voltage U_i, this voltage must be higher than the breakdown voltage in con-

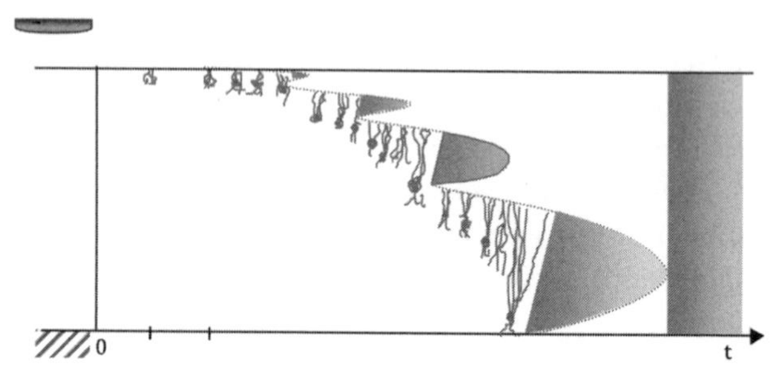

FIGURE A.4: DEVELOPMENT OF A NEGATIVE DISCHARGE (NEGATIVE ROD, PLANE GROUNDED ELECTRODE)

tinuous current U_s. The formation time lag decreases with the increasing of the voltage gradient $U_i - U_s$.

In smaller gaps (decimeters or so), the streamer to spark transition is straightforward when the streamer crosses the gap and reaches, for example, the plane grounded electrode.

In long gaps, the first step in the discharge development is a primary corona effect, called *first corona*, looking like streamer burst (corona streamers) coming from the high-voltage electrode. The second step is the development of a highly conductive channel, called *leader*, also coming from the high-voltage electrode. As a third step, the leader spreads out thanks to corona discharges coming from its head and moving to the grounded electrode. Finally, the *return stroke* is initiated when the streamers originating from the corona effect at least one leader head reach the grounded electrode.

In a "negative rod—grounded plane" gap, a first corona and a dark region (similar to a "positive rod—grounded plane") are observed. However, the luminous intensity and the length of the corona streamers are much weaker.

After the dark period, a unique process called *pilot system* with short duration brilliant dots appears. From these, both polarities streamers develop in two opposite directions (see big dots in figure A.4).

The interaction between positive streamers so generated and first corona streamers leads to the generation of a *negative leader* coming from the cathode. The process is repeated with regular time intervals during the leader propagation by successive steps.

When the leader is completely developed, breakdown occurs starting with the *return stroke* when the first corona streamers reach the plane. This phenomenon generates a sharp reillumination of these corona streamers from the plane to the leader head. The return stroke is followed by the complete discharge which is an electric arc bridging the two electrodes.

General Reference

V. Cooray, *The Lightning Flash*, IEE, Power & Energy Series 34, UK, 2003.

GLOSSARY

Air-termination system: part of an external (structural) lightning protection system using metallic elements such as rods, mesh conductors, or catenary wires intended to intercept lightning flashes.

Amplitude I of the lightning current: peak value I of the lightning current i.

Anvil: upper part of a cumulonimbus, typically containing a positive charge, and often with a screening layer of negative charge at the cloud boundaries.

Ball lightning: luminous formation with a lifetime of a few seconds . . . still a puzzle in physics!

Capacitance: see frame in chapter 4.

Coordinated SPD protection: set of surge protective devices properly selected, coordinated, and installed to reduce failures of electrical and electronic systems.

Cumulonimbus: (Cb) thunderstorm clouds, with great vertical extent, sometimes reaching 50,000 to 60,000 feet. They can have vigorous convective updrafts (sometimes in excess 50 knots); they are characterized by liquid water at low levels and ice crystals and super-cooled liquid water (liquid water at temperatures well below 0 degrees Celsius) at high levels.

Corona effect: see appendix.

Dart leader: see **leader.**

Dielectric strength: see frame in chapter 3.

Down-conductor system: part of an external (structural) lightning protection system intended to conduct lightning current safely from the air-termination system to the earth-termination system.

Downward flash: lightning flash initiated by a downward leader from cloud to earth. A downward flash consists of a first return stroke and possibly several subsequent return strokes. Steady currents known as continuing currents may follow impulsive current components.

Earthing electrode: (in U.S., grounding electrode) part of the earth-termination system, which provides direct electrical contact with the earth and disperses the lightning current into the earth.

Earth-termination system: part of an external (structural) lightning protection system, which is intended to conduct and disperse lightning current into the earth.

Electrical conductivity: see frame in chapter 4.

Electrical discharge in air: see appendix.

Electrical potential energy: see frame at chapter 4.

Electrical system: power supply system (mains).

Electric charge: see frame in chapter 2.

Electric current: see frame at chapter 4.

Electric field: see frame in chapter 3.

Electric field lines and **electric spectrum:** see frame in chapter 3.

Electric force: see frame in chapter 2.

Electric induction: see frame in chapter 2.

Electric potential or **voltage:** see frame in chapter 3.

Electric power: see frame in chapter 4.

Electrization: current flowing, not necessarily deadly, through the human (or animal) body.

Electrocution: deadly electrization by ventricular fibrillation (or a systole) leading to an irreversible cardio-respiratory arrest.

Electrogeometric model or **rolling sphere model:** see chapter 8.

Electronic system: system incorporating sensitive electronic components such as communication equipment, computer, control and instrumentation systems, radio systems, power electronic installations.

External conductive parts: extended metal items entering or leaving the structure to be protected such as pipes, cable metallic elements, and metal ducts, which may carry a part of the lightning current.

External (structural) lightning protection system: part of the lightning protection system consisting of an air-termination system, a down-conductor system, and an earth-termination system.

Failure of electrical and electronic systems: permanent damage to electrical and electronic systems due to the electromagnetic effects of lightning.

Faraday cage: metallic enclosure (conductors, grids) which ensures that lightning current is constrained to its exterior.

Flash charge: time integral of the lightning current for the entire lightning flash duration.

Flash duration: for a Cloud-to-ground flash, the time for which the lightning current flows at the point of strike (including no-current intervals between strokes). For Intra-cloud flashes, duration may be a more subjective characteristic.

Fulgurites: glassy tubes made when lightning current flows in sandy soil.

Induced discharge: see chapter 7.

Internal systems: electrical and electronic systems.

Ionosphere: upper region of the atmosphere, situated between the neutropause and an altitude of about 500 km (ionopause), where the charged particle concentration is quite larger than within the neutrosphere.

Keraunic level or **number of thunderstorm days T_d:** annual number of days when thunder has been heard in a given place.

Lateral discharge: see chapter 7.

Leader: first stage of a lightning stroke corresponding to the formation of an ionized channel most often between cloud and ground. A **downward leader** propagates by steps from cloud to ground, it is generally negative (90% in summer), and is called a stepped leader. An **upward leader** is initiated by grounded objects and propagates upward to cloud or to a downward leader tip. In a **positive leader**, the propagation is generally continuous (not by steps). Subsequent leaders following the first stepped leader are called **dart** or **dart-stepped leaders**. Dart leaders propagate continuously. Subsequent leaders may deflect from the previously formed channel and become stepped leaders.

Lightning current i: current flowing at the point of strike.

Lightning electromagnetic impulse (LEMP): electromagnetic effects of lightning current. It includes conducted surges as well as radiated impulse electromagnetic field effects.

Lightning equipotential bonding: bonding to the lightning protection system of separated metallic parts by direct conductive connections or via surge protective devices (SPDs) to reduce potential differences caused by lightning current.

Lightning flash density N_g: annual number of lightning strikes to earth per unit area ($km^{-2}.year^{-1}$).

Lightning flash near an object: lightning flash striking close enough to an object to be protected that it may cause dangerous over-voltages.

Lightning flash to an object: lightning flash striking an object to be protected.

Lightning flash to earth: one of the lightning discharges of atmospheric origin, between earth and cloud, consisting of one or more strokes.

Lightning protection level (LPL): number related to a set of lightning current parameter values relevant to the probability that the associated maximum and minimum design values will not be exceeded in naturally occurring lightning. Lightning protection level is used to design protection measures according to the relevant set of lightning current parameters.

Lightning protection system (LPS): complete system used to reduce physical damage due to lightning flashes to a structure. It consists of both external and internal lightning protection systems.

Lightning protection zone (LPZ): zone where the lightning electromagnetic environment is defined and controlled. Within an LPZ, the electromagnetic effects of the lightning current can be reduced.

Lightning rod: see **external lightning protection system.**

Lightning stroke: single electrical discharge in a lightning flash to earth.

Long stroke: part of the lightning flash, which corresponds to a continuing current. The duration of continuing current is typically more than 2 ms and less than 1 s.

Long stroke charge: time integral of the lightning current in a long stroke.

Magnetic shield: closed, metallic, grid-like, or continuous screen enveloping the object to be protected or part of it, used to reduce failures of electrical and electronic systems.

Measuring connection: connection that can be dismantled between the down-conductor and the corresponding earthing electrode, to allow the measuring of the electrical resistance of earthing electrode.

Meshed cage: Faraday cage with large meshes (5 to 20 m dimensions) completely surrounding a structure to be protected.

Mesopause: upper limit of the mesosphere, situated at an altitude of about 85 km.

Mesosphere: middle region of the atmosphere above the stratopause, where the temperature decreases when the altitude increases, to reach a minimum at the mesopause.

Multiple strokes: lightning flash consisting on average of 3 to 5 strokes with typical time intervals between them of about 60 ms. Events having up to a few dozen strokes with intervals between them ranging from 10 ms to 250 ms have been reported.

Neutropause: upper limit of the neutrosphere, situated at an altitude of about 60 km.

Neutrosphere: low atmosphere region situated between ground and neutropause where the charged particle concentration is insignificant.

Object to be protected: structure or service to be protected against the effects of lightning.

Peak value I: maximum value of the lightning current i (see amplitude).

Physical damage: damage to a structure (or to its contents) or to a service due to mechanical, thermal, chemical, and explosive effects of lightning.

Plasma: ionized medium where the positive charge is macroscopically in equilibrium with the negative charge; it is sometimes called the *fourth state of matter*.

Point of strike: (ground-strike point) point where a lightning flash strikes the earth or protruding object (e.g., structure, lightning protection system, service, tree). A lightning flash may have more than one point of strike (multi-channel flash).

Protection measures: measures to be adopted in the object to be protected to reduce the lightning risk.

Resistivity: see frame in chapter 4.

Return stroke: lightning process following the junction between a downward leader coming from the thundercloud with the upward connecting leader coming from ground; a high-current surge propagating from ground to cloud.

Ring conductor: conductor forming a loop around the structure and interconnecting the down-conductors for distribution of lightning current among them.

Ring earthing electrode: earthing electrode forming a closed loop around the structure below or on the surface of the earth.

Risk R: ratio of value of probable average annual loss (humans and goods) due to lightning, relative to the total value (humans and goods) of the object to be protected.

Saint Elmo's fire: another name for corona effect, particularly observed on ship masts or other high sharp objects.

Service to be protected: service connected to a structure for which protection is required against the effects of lightning.

Shielding wire: metallic wire used to reduce physical damage due to lightning flashes to a service. Shield wires (overhead ground wires) are used above phase conductors on high-voltage transmission lines to intercept the lightning current and conduct it to ground through neighboring metallic towers.

Short stroke: part of the lightning flash, which corresponds to an impulse current. This current has a duration typically less than 2 ms.

Short stroke charge: time integral of the lightning current in a short stroke.

Specific energy W/R: time integral of the square of the lightning current for the entire flash duration. It represents the energy dissipated by the lightning current in a unit resistance (1 Ω).

Specific energy of short stroke current: time integral of the square of the lightning current for the duration of a short stroke. The specific energy for a long stroke current is assumed to be negligible.

Step voltage: see chapter 7.

Stratopause: upper limit of the stratosphere, situated at about 45 km from ground.

Stratosphere: region of the atmosphere situated between the tropopause and the stratopause where, when altitude increases, the temperature first decreases a little, then remains constant (up to about 25 km) and increases because of the absorption of solar UV energy by ozone.

Streamer: see appendix.

Striking distance: distance between the head of the downward leader and the point of strike at the time of initiation of an upward connecting leader.

Structure to be protected: structure for which protection is required against the effects of lightning in accordance with the international standard IEC 62305. A structure to be protected may be a part of a larger structure.

Surge protective device (SPD): device intended to limit transient over-voltages and divert surge currents. It contains at least one non-linear component.

Thermosphere: highest region of the atmosphere, beyond the mesopause, where the temperature constantly increases to reach 1,000 K (1273 °C) at the altitude of 750 km.

Thunder: set of elastic waves generated in air by high-energy acoustic pressure waves along the lightning channel axis, transmitted and then refracted by different objects along their path to the receptor.

Tolerable risk R_T: maximum value of the risk which can be tolerated for the object to be protected.

Touch voltage: see chapter 7.

Tropopause: upper limit of the troposphere, situated at an altitude between 10 and 17 km.

Troposphere: lower part of the atmosphere, between ground and tropopause, where the temperature decreases when the altitude increases.

Upward flash: lightning flash initiated by an upward leader from an earthed structure to cloud. An upward flash consists of a first long stroke (initial-stage current) with or without superimposed pulses. The initial-stage current may be followed by subsequent strokes, similar to those in downward flashes.

Voltage: see **electric potential** (frame in chapter 3).